THE LOST MILLENNIUM

THE *Lost*

FLORIN DIACU

Millennium

History's Timetables Under Siege

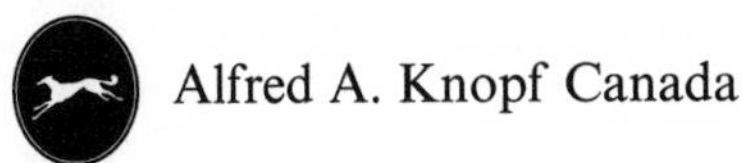

Alfred A. Knopf Canada

PUBLISHED BY ALFRED A. KNOPF CANADA

Published in 2005 by Alfred A. Knopf Canada, a division of Random House of Canada Limited. Distributed by Random House of Canada Limited, Toronto.

Knopf Canada and colophon are trademarks.

www.randomhouse.ca

Figures 2.3, 6.1, 6.2, 6.3, 6.4, 6.5, 6.6, and 2.4 are from the forthcoming book, Fomenko, Anatoli, T., Tatiana N. Fomenko, Wieslaw Z. Krawcewicz, and Gleb V. Nosovski. "Mysteries of Egyptian Zodiacs." Unpublished manuscript, Edmonton, 2003, and are reprinted with permission.

LIBRARY AND ARCHIVES CANADA CATALOGUING IN PUBLICATIONS

Diacu, Florin N.

The lost millennium : history's time tables under siege / Florin Diacu.

ISBN-13: 978-0-676-97657-1
ISBN-10: 0-676-97657-3

1. Calendar—History. 2. Chronology, Historical. I. Title.

CE6.D52 2006 529'.3 C2005-901155-6

Text design: CS Richardson

First Edition

Printed and bound in the United States of America

10 9 8 7 6 5 4 3 2 1

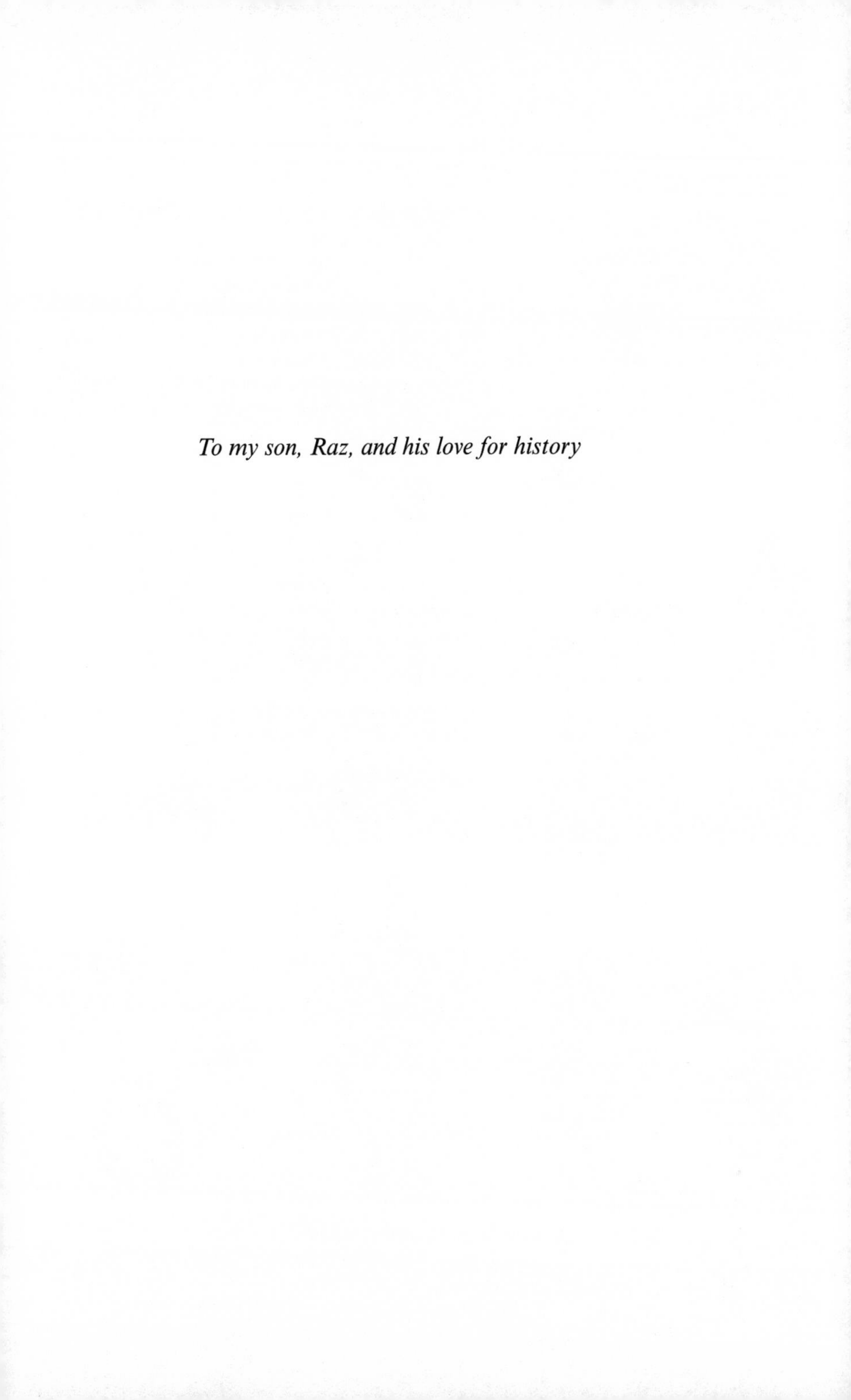

To my son, Raz, and his love for history

People believe in the truth of all that seems to be strongly believed in.

FRIEDRICH NIETZSCHE

We should try to love the questions themselves, like locked rooms and like books that are written in a very foreign tongue.

RAINER MARIA RILKE

Contents

INTRODUCTION

Where Did the Time Go?

Those whose chronology is confused cannot give a true account of history.

TATIAN

Mexicans call Cuernavaca "the city of the eternal spring." In the Tepozteco Valley, where the city rests, the mornings are clear, the afternoons turn hazy, and the evenings are blessed with a tropical rain.

I spent a week in September 1994 a few miles from the city, in the hacienda-style resort of Cocoyoc. The place would have resembled the Garden of Eden were it not for the volcanic mountain Popocatépetl, which—though too far away to pose an imminent threat—loomed in the distance, rings of smoke hanging above its icy cone. A conference had brought together mathematicians from three continents. All week we had listened to lectures, solved problems, discussed ideas, and learned new techniques to help us keep up with developments in our field.

On the last day of the conference, I was having lunch with fellow mathematicians Tudor Ratiu and Ernesto Pérez-Chavela. Like me, Tudor had been born and raised in Romania. Nine years my senior, he now taught at the University of California in Santa Cruz. Ernesto, a young professor at the Universidad Autónoma Metropolitana-Iztapalapa in Mexico City, was a co-organizer of the Cocoyoc meeting.

During lunch, Ernesto told us the story of Cocoyoc. In the local dialect, *cocoyoc* means coyote, an animal often seen in the area centuries ago. The resort, endowed with swimming pools, tennis courts, and a golf course, had once been a hacienda and apparently had been founded almost five centuries ago by the Spanish conquistador Hernando Cortés. Ernesto's account of Cocoyoc's origin sounded like a legend. It might have been true, but it made us wonder how much fiction finds its way into the history books.

"It's a fascinating subject," Tudor said. "It reminds me of a Russian colleague, Anatoli Fomenko, who thinks that a lot of the 'historical record' is fiction. So he's researching history with mathematical tools."

Ernesto looked surprised, and I must have too, for Tudor asked if we knew about Fomenko. We hadn't heard of him before, but the idea of applying mathematics to the study of history seemed interesting enough. My knowledge of such applications didn't go beyond understanding the simple differential equation that explains carbon dating.

"He's from the University of Moscow," Tudor explained, "and is quite active in several fields of mathematics. Something of a polymath. I met him in Berkeley a few years ago. His work in chronology has convinced him that the Middle Ages never happened. Apparently the authorities who fixed the dates misinterpreted the ancient documents, and their mistakes have been perpetuated ever since. Fomenko believes that the history of humankind is about a thousand years shorter than we think."

"He can't be serious," I said.

History has been an interest of mine since I was eleven. In my early teenage years I wanted to become an archaeologist, to discover and explore ancient ruins and unravel the mystery of lost kingdoms. I was fascinated with the idea of digging the earth and finding traces of dead civilizations. The curiosity I felt for antiquity was fuelled by the books I had read about Mesopotamia and Egypt, the Hittites of Asia Minor, the Hebrew and the Minoan-Mycenaean civilizations, early India, China, and Eurasia, the Assyrian Empire, Dacia, Thrace, and the Greek and Roman worlds.

But my gift for the exact sciences and success in mathematics competitions steered me in a different direction. Nevertheless, my interest in antiquity survived, and I kept up my reading in ancient history, watched documentaries, and continued to learn new things about the distant past of humankind. So, not surprisingly, my first reaction to Fomenko's claim was total disbelief.

"He's very serious," Tudor said, "but don't ask me why. If I remember well, it's not only the Middle Ages. He thinks that several shorter periods, which add up to a thousand years, have been created by mistakes in the dating process."

"A millennium that lost its way in history?" Ernesto asked.

"Something like that."

"Has he published anything about it?" I inquired.

"Plenty. I got a paper from him last week, a day or two before leaving Santa Cruz."

"What about?" I replied.

"It's an examination of ancient and medieval dynasties. He argues that many of them overlap instead of being successive."

"That's hard to believe," I said. "Real historians must have thought about those things."

"I'd give him the benefit of the doubt," Ernesto said. "Think of Einstein or Newton or Darwin. They were unknown in their field once, but they proved everyone wrong. That's how scientific revolutions happen."

"Perhaps you're right," I said. "I have no experience with chronology. Still, this sounds incredible." Then, turning to Tudor, I asked: "Is Fomenko trustworthy?"

"I don't know him well, but he's a brilliant mathematician. He has written a dozen books and more than a hundred articles—excellent, as far as my field is concerned. I've also heard that he's just been elected to the Russian Academy of Sciences. It's a highly respected institution."

Tudor and I had spent hours talking together that week in Cocoyoc. He's a skilled mathematician, and I trusted his judgment. If he didn't dismiss Fomenko's claims from the outset, it meant I had to keep an open mind. But I would have liked to see the arguments.

"What do *you* think?" I asked. "Is he right?"

"He's not bluffing, but I have no idea if he's right. Other people must agree with him, otherwise he wouldn't be able to publish this stuff in serious journals."

"Has he written any books on chronology?" Ernesto asked.

"Yes, in Russian. But—if I remember correctly—an English translation is either about to come out or is in print already."

I made a mental note to track it down once I returned home.

"From what I remember of the history I learned in school," Ernesto said, "the Middle Ages are not well documented."

"This is definitely true for the history of Romania," I said. "The Romans conquered Dacia in AD 106, then mixed with the Dacians and imposed their language and culture. But in 271 they withdrew their legions and moved them south of the Danube, which was a good shield against barbarian attacks. From then until the twelfth century we know only about the kings Gelu, Glad, and Menumorut, who reigned over some parts of Transylvania, and whom the Hungarian rulers fought when invading the region. But aside from these details, more than eight hundred years of Romanian history are unaccounted for."

"What do the historians say?" Ernesto said.

"It depends on who you ask," I replied. "Romanians claim it took a thousand years of mixing between the Slavic invaders and the local Dacian-Roman people before the Romanian nation was born. But Hungarians argue that nobody lived there."

"There is no forest without beasts," Ernesto said. "Why would people stay away from fertile land? Somebody must have been there. But where are the signs of their occupation?"

Ernesto was right. Such a gap didn't make sense. The people who lived in Transylvania were civilized enough to leave traces. Why hadn't I thought of that? Perhaps because, like everyone else, I had learned those things in childhood and never doubted my teachers. I reflected for a moment on how hard it is to break free from the "truths" acquired at an early age.

"Fomenko's theory might explain this information gap and perhaps more," Tudor said. "One thing I always found hard to accept is that there was no progress in the Middle

Ages for almost a thousand years. To me, this contradicts the questing of the human spirit. Can you believe that nobody wrote books, created art, or developed science in Europe from the fall of Rome until the Renaissance?"

"I read once about a star catalogue attributed to Ptolemy of Alexandria," Ernesto said. "The trouble is, the sky configuration recorded there appeared only a thousand years after him."

"So, another point for Fomenko," I said. "But let's think of arguments against him. What about carbon dating?"

"I think it can give large errors," Tudor said. "Moreover, it was calibrated to fit what historians believe is correct data."

"What about the Shroud of Turin?" I said. "Three independent tests placed it in the thirteenth or fourteenth century. Those experts can't be all wrong."

"I think those tests are fine," Tudor said. "The method seems to be unsuitable for much older objects. But again, I don't speak as an expert."

"So for the period we're interested in," I said, "we could rely on carbon dating. That would be an argument against Fomenko."

"Not necessarily," Ernesto replied. "What if the Shroud still belonged to Jesus, and he was born less than a thousand years ago?"

"But there must be other objects that were dated correctly," I said. "It's just that they don't make the news."

"That's probably true," Ernesto said. "In fact, I'm thinking of another argument against Fomenko: the time span between the introduction of the Julian and the Gregorian calendars. Because Pope Gregory XIII reformed the calendar in the sixteenth century by deleting the ten days that had accumulated from the mismatch between the astronomical and the calendrical year, we can date the Roman period pretty accurately."

"Only if you assume that Pope Gregory's astronomers were correct," Tudor said. "Fomenko thinks they were not. The problems of chronology are far from easy. Perhaps only a multidisciplinary group can understand them. We've mentioned astronomy, carbon dating, calendars, and I'm sure other disciplines can also get involved. Linguistics might be a candidate. By comparing, say, Dutch and German—looking at how much the vocabulary has changed, for example—you can tell when these languages began to split."

We continued to chat about chronology, but none of us was an expert in any of the subjects related to it. The best we could do was to ask questions that came to mind and guess the answers. Still, we knew we could be wrong. Science and mathematics have so many examples of intuitive "truths" that are false. The "flatness" of the Earth is a typical "truth" of earlier times.

Our discussion ended with that lunch. We went to the last talks of the conference, and our minds turned back to mathematics. But the idea of a lost millennium made a strong impression on me. I stored every detail of our chat in my memory. For my two friends, our conversation had been small talk: years later, Tudor would remember it vaguely, and Ernesto not at all. They never took the issue very seriously.

Back home in Canada, I tried to find Fomenko's book. I had only two easy choices: the University of Victoria Library and the campus bookstore, and neither had it on the shelves. Years later, I learned that while I was making my first attempt to find that book, the English edition was published.

In 1994, disappointed with my search, discovering that Fomenko's articles were written in Russian—a language I wish I knew more about—and swept up with my academic duties, I postponed any further effort to find an English translation. I focused on my research in celestial mechanics, wrote a book about chaos theory, secured tenure, and got promoted. I enjoyed my work and knew I should not allow myself to be distracted from it.

Time passed. In 2000 I became a full professor. From then on I could diversify my research interests without worrying about my list of publications. Chronology again came intermittently to mind. Then, on September 16, six years after hearing Tudor talk about Fomenko's theory, I found a copy of *Saturday Night* magazine inside the morning newspaper. As I leafed through its pages, the picture of a middle-aged man standing in an elevator caught my attention. He had a large forehead and an unkempt beard, was dressed in a dark shirt and a pair of jeans held up by suspenders, and had two books in his hand. Thoughtful, unconcerned with his appearance, he looked the typical North American professor. Then I read the article's title, "Time Warp," and the first sentence: "You might think it's the year 2000, but a group of prominent Russian mathematicians is arguing that history is all wrong, and it's actually 936 AD."

Breakfast had to wait until I finished reading the article. Timothy Taylor, its author, had clearly done his homework. The biggest surprise, however, was the man in the picture, Wieslaw Krawcewicz, a mathematics professor at the University of Alberta, in Edmonton, with whom I served as editor of a high school magazine sponsored by the Pacific Institute for the Mathematical Sciences (PIMS). Only a few months before, I had participated in a meeting of the PIMS board and executive, which approved the funding for *Pi in the Sky*—as we decided to call the new publication. Wieslaw and I had corresponded for some time, but we had not yet met.

Krawcewicz was at the centre of the controversy discussed in the article. He had invited a Russian mathematician, Gleb Nosovski, to his university to give a talk about chronology. Nosovski had been a student and long-time collaborator of Anatoli Fomenko, and he worked as a senior researcher at the Laboratory of Computer Methods in Natural and Human Sciences at the University of Moscow. When he learned about this event, Taylor—who lived in Vancouver—went to Edmonton to attend it. He found a room packed with faculty and students, as well as many members of the public.

Though he was busy taking notes and trying to understand Nosovski's English, Taylor noticed that listeners reacted in different ways. The man in a tweed jacket sitting in front of him was growing restless, shaking his head from time to time, obviously irritated with Nosovski's claims, while Krawcewicz, who sat at the end of the front row, was surveying the audience, clearly enjoying what he saw.

In the first part of his talk, Nosovski criticized the traditional chronology founded by the sixteenth- and seventeenth-century scholars Joseph Scaliger and Dionysius Petavius. He offered astronomical explanations for why the Peloponnesian War between the Greek city-states of Athens and Sparta couldn't have taken place in the fifth century BC but, rather, must have occurred in the eleventh or the twelfth century AD, and why the eclipse described in Livy's history of Rome must have happened in the tenth century AD instead of the second century BC. He compared the dynasties of kings considered to have lived more than a millennium apart and outlined statistical arguments why many of them had to be duplications. He mentioned a book by Isaac Newton which claimed that the chronology of ancient Greece was too long by about three centuries. He also analyzed several

Egyptian horoscopes and concluded that they showed configurations of the sky which appeared much later than the dates attributed to them.

In the second part of his talk, Nosovski presented a new chronology in which the succession of the main historical dates agreed with his mathematical evidence. He placed most ancient events seven to ten centuries closer to our time than tradition did, but admitted that this new system was still under construction.

Some people in the audience grew impatient. The man sitting in front of Taylor was described in the article as being visibly agitated. Another listener tried to interrupt the speaker a couple of times.

Nosovski closed his talk with a transparency that revealed the results of his computations regarding Jesus' year of birth. One man in the audience said this couldn't possibly be true, but the applause at the lecture's end silenced him. While some listeners raised their hands to ask questions, and others took the opportunity to leave the room, Taylor noticed how deeply divided the audience had grown.

For him, this lecture was but the beginning. Like any good journalist, Taylor didn't content himself with only one point of view. He needed to hear what historians had to say. Christopher Mackay, an associate professor from the Department of History and Classics at the University of Alberta, agreed to speak with him. Holding a degree from Harvard in classical philology, Mackay was an expert in Greek and Latin literature, in Roman history and law, and in Latin epigraphy—the science that deals with deciphering and interpreting inscriptions. In his third-year course on early Roman history, he taught a section about the chronology of Rome.

Mackay considered the claim about Jesus absurd. In his opinion, it would mean that the First Ecumenical Council of Nicaea preceded the birth of Jesus by about six centuries. This gathering, which had taken place in the ancient town of Nicaea (now Iznik, in northwestern Turkey), is best known for having established the earliest dogmatic statement of the Orthodox Church. The Roman emperor Constantine the Great convened the council in AD 325 in an attempt to settle the controversy raised by Arianism over the nature of the Christian Trinity: the Father, the Son, and the Holy Ghost.

The council formalized the Nicaean Creed, according to which God the Father and God the Son are eternal, and declared heretic the Arian belief in a Christ inferior to the Father. Arius, the Christian leader who had spread the belief in a lesser Christ, was excommunicated. The council also imposed a code of ethics, discipline, status, and jurisdiction for the clergy and established the Easter dates.

In Mackay's reasoning, the Russian mathematician used certain fragments of information, among them the fact that Jesus died when he was thirty-one years old, that the resurrection took place on March 25 (which was a Sunday), that the Passover—an annual Jewish holiday in memory of the Hebrew slaves' escape from Egypt—fell on March 24, and that the Easter dates are calculated using an old church text called the Easter Book. Mackay said the dates and the Easter Book were medieval additions, and therefore unreliable. Since Nosovski's calculations were not based on the Bible, they did not have the value of proof.

Then Taylor asked about Isaac Newton. Just before his death in 1727, the English scientist had completed a book entitled *The Chronology of Ancient Kingdoms Amended*, which was published a year later. Newton had been interested in history and chronology since his student years, as his unpublished manuscripts show, and he always felt that historical chronology was flawed—particularly in relation to the history of ancient Greece. His main argument rested on the incorrect dating of the Argonautic Expedition.

Mackay dismissed Newton's work in this area as he had Nosovski's, explaining that scientists in Newton's time had no knowledge of the ancient Egyptian, Babylonian, or Sumerian languages. He said Newton's conclusion was as absurd as Euclid's interpretation of Einstein would be. This analogy had some force, but it didn't convince me.

If Newton were not to be trusted because of his era, why would his near contemporaries have more credibility? As perhaps the most respected of all scientists, Newton couldn't be dismissed in a sentence. I knew I had to read his book carefully and find out what other people thought about it, for I couldn't imagine that it had passed unnoticed for almost three centuries.

Taylor also mentioned the work of the Russian polymath Nicolai Morozov, who, between 1924 and 1932, had published a seven-volume work entitled *Christ: The History of Human Culture from the Standpoint of the Natural Sciences*, in which he examined chronology using mathematical, astronomical, linguistic, philological, and geological arguments. He was the first to suspect that ancient history, as we know it, is about a millennium too long. In his Edmonton talk, Nosovski had often referred to Morozov.

But Mackay was not impressed with Morozov either. He told Taylor that Steven Hijmans, an archaeologist at the University of Alberta, had found some data irregularities in Morozov's work, as Krawcewicz presented them in one of his articles on chronology. Besides, Mackay raised doubts about Morozov's character, claiming that the Russian had been a rich kid, a thug, and possibly a Bolshevik because he survived under Stalin until 1946. If his character was that flawed, could his conclusions be trusted?

I was surprised to hear this argument coming from a scholar who studies people in the context of their society. Mackay seemed to know little about the Soviet Union. As one who lived for three decades in a similar society, I saw how those who confronted the authorities had perished. Still, the totalitarian regime had not stopped the Soviet Union from developing in the areas of science and technology. Many top researchers were members of the Communist Party, yet that didn't mean they believed in Communism. Dismissing Morozov as a rich kid, a thug, and possibly a Bolshevik was unconvincing. Morozov's work was another source I had to investigate.

Taylor also wanted to hear Mackay's opinion on the chronology of Egypt. Nosovski had mentioned that some horoscopes found in tombs showed configurations of the sky that appeared more than a millennium after the pharaohs' deaths. To this, Mackay answered that he preferred to stick with the Egyptologists, dismissing the possibility that a single horoscope could contradict two centuries of research.

This response didn't satisfy me either. Invoking an authority is not a proof. I would have preferred to learn from Mackay whether those horoscopes might be just artwork with fictional planetary configurations. Still, he brought some interesting arguments against the new dating of the Peloponnesian War to the eleventh or twelfth century AD, and of Livy's eclipse to the tenth century AD. His objection was not to the shift of events in time but to the reversing of their order, which put the effect in front of the cause.

Mackay began with the Battle of Pydna (168 BC) between Macedonia and Rome, named after a town near the shore of Thessaloniki's gulf. Traditional chronology claims that the Peloponnesian War preceded the Battle of Pydna by more than two and a half centuries. Mackay objected not to Nosovski's putting the battle a millennium later, but to his placing it after the Peloponnesian War.

During that time, mainland Greece and western Asia Minor were the heartland of the Greek world. Eighty years after the war's end, Mackay points out, Alexander the Great conquered the Persian Empire, Greeks spread throughout the Near East, and three major Greek monarchies were established: the Seleucid, the Antigonid, and the Ptolemaic. The Antigonid dynasty was brought to an end when the Romans defeated it at the Battle of Pydna. That Roman victory assumes the existence of Alexander the Great. Without him, Mackay concludes, there would have been no Antigonid dynasty and no Battle of Pydna.

Of course, Taylor still had other interviews to conduct. He approached Krawcewicz, who proved to be on Nosovski's side. In one of his articles, Krawcewicz had characterized the dating techniques used by historians and archaeologists, including the radiocarbon method, as "highly subjective and based on presumptive evidence." Since these methods were a key factor for invalidating the Russians' objections to traditional chronology, I had to learn more about them and form my own opinion on how precise they might be.

Taylor consulted two other mathematicians, Jacques Carriere and Jack Macki, both from the University of Alberta, as well as the astronomer Roger Sinnott—an editor of *Sky & Telescope* magazine. Macki and Carrière agreed that the mathematics of Nosovski and Fomenko was sound, though they found it difficult to use for drawing drastic conclusions about chronology. Sinnott also doubted the historical interpretations, but confirmed the correctness of the astronomical data the Russians had employed.

One colourful presence in Taylor's article was the former world chess champion Garry Kasparov, whose games against IBM's Deep Blue supercomputer in the 1990s had made headlines all over the world. Though neither a research mathematician nor a professional historian, Kasparov counted ancient history among his hobbies. Taylor had a chance to interview Kasparov and learn what the grand chess master thought about chronology.

Kasparov had met Fomenko a few years earlier and he found that the Russian mathematician's conclusions confirmed some of his own concerns. Kasparov has an excellent memory, one trained to recognize patterns, and he had discovered a number of absurdities that historians couldn't explain. His main problem period was the relatively blank era of the Dark Ages—an issue Tudor had also raised during our discussion in Cocoyoc. Kasparov could not accept the commonly held belief that, after the collapse of the western Roman Empire, art and science had died and had then taken a thousand years to recover. It seemed illogical that the many Roman citizens who moved to Constantinople had left all the scientific knowledge of Rome behind. How could the principles of mapping and the science of ballistics vanish, when anything that has military significance is protected and encouraged by any state, no matter who the leader?

"Time Warp" was an objective and well-written article. Timothy Taylor had collected the opinions of several specialists from various fields and had done his best to make this interesting and controversial issue accessible to the public. In the spring of 2002 I met him at the University of Victoria, where he read from his successful first novel,

Stanley Park. Talking to him after the reading, I learned that "Time Warp" had won a National Magazine Award.

My own experience as a researcher, however, made me understand that the only well-considered point of view presented in "Time Warp" was that of Fomenko, Nosovski, and Krawcewicz. Though they had degrees in history, mathematics, or astronomy, none of the other people interviewed had even read the work of the Russians, let alone researched it. A talk has only informal value; it provides no insight into the arguments accumulated after years or decades of thought. Only an expert who has spent weeks or months trying to understand the facts can form an opinion on whether claims are true or false. And no such opinion appeared in Taylor's article.

But "Time Warp" revived my interest in chronology. It made me ask new questions and develop a research plan. The same day, I started searching library and Internet resources, looked for Fomenko's works, tried to find books that explained how traditional chronology came into being, and sent an email to Wieslaw Krawcewicz asking for more details. The seed sown six years earlier had finally sprouted.

The soil had been fertilized not only by my interest in antiquity but also by my teenage experience. While growing up in Romania, I had witnessed how Nicolae Ceausescu's regime rewrote history. Convenient aspects were emphasized, inconvenient details omitted, and the past kept on changing. This had made me reflect on the fragility of history, on how ideology and power can bend the truth. How far could this distortion go? Though I doubted it had affected ancient chronology, I had to investigate.

I planned to do this work without neglecting my mathematics research and teaching. Chronology would become part of my "history hobby," nothing more. In the next few years I read the literature; got in touch with Nosovski, Fomenko, and Kasparov; talked to various astronomers, archaeologists, chemists, physicists, and mathematicians; and had discussions with several historians who specialize in chronology, trying to grasp the many points of view in the field.

This fascinating experience led me in unexpected directions. I had to understand how Scaliger and Petavius thought, what Newton's main objections were, who Morozov was and why he took the risks he did, how the radiocarbon method works and whether it's reliable enough for the purposes of ancient and medieval chronology, how far we can base our conclusions on astronomical computations and recorded data, and why statistics matters to historical research. When I started, it had never occurred to me that I would learn about so many achievements in such a multitude of human activities.

In September 2002 I received a visit from Donald Saari, a colleague and friend from the University of California at Irvine. Don is not only a renowned mathematician and economist but also a man with a wide cultural perspective. We had dinner together, and I told him what I had learned about chronology. After listening for half an hour, he said: "This sounds like a detective story. Share it with people, write a book about it."

The idea didn't appeal to me at first. I wanted to continue my adventure, not simply recount it. But Don's suggestion stuck with me. Putting my conclusions on paper would help me focus my research and give me a better understanding of the facts. Soon I found myself outlining the chapters, and one day I started to write.

This book describes my journey in the field of historical chronology. The people who appear in these pages—whom I met in person, over the phone, or through their writing—

belong to different cultures and generations, speak different languages, and have different backgrounds, research methods, and ways of attacking problems. They often disagree about the interpretation of historical documents, but all of them strive towards a common goal: to explain the distant past of humankind.

Perhaps Don was right: the issues treated here and the questions they pose might make these pages read like a detective story. But beyond mystery, this subject draws attention because of the debate and the excitement it stirs. These ingredients have stimulated my curiosity and have pushed me further and further in search of the truth.

I hope you'll enjoy this adventure as much as I did.

PART ONE

The Challenges of Historical Chronology

CHAPTER I

Catastrophes and Chaos

No testimony is sufficient to establish a miracle, unless the testimony is of such a kind that its falsehood would be more miraculous than the fact it endeavours to establish.

DAVID HUME

In January 1950 *Harper's* magazine published an article that generated an immediate controversy. "The Day the Sun Stood Still," by Eric Larrabee, marked the beginning of an unusual dispute. The article outlined the ideas of a fifty-four-year-old Russian physician, Immanuel Velikovsky, who maintained that "the experts" had got the histories of ancient Israel and Egypt all wrong. He set out to prove it in two new books: *Worlds in Collision* and *Ages in Chaos.*

No other reformer of chronology ever had such an impact on public opinion as Velikovsky. The Russian doctor captured the media's attention for months, leading to debates that often turned nasty. This polarized attitude confused most observers. Fed up with the insults exchanged on

TV or in the press, they wanted the answer to only one question: Is Velikovsky's theory true?

WORLDS IN COLLISION

As a student, Velikovsky had excelled in languages and mathematics, graduating from high school with a gold medal—the equivalent of an A+ in all subjects. Receiving his medical degree from the University of Moscow in 1921, he moved to Berlin to edit the scholarly journal *Scripta Universitatis*, for which Albert Einstein took charge of the physics/mathematics volume. After 1924, Velikovsky successively practised general medicine in Jerusalem and psychoanalysis in Haifa and Tel Aviv. The outbreak of the Second World War found him in New York, where he began writing a book about Sigmund Freud.

In April 1940, while reading Freud's *Moses and Monotheism*, Velikovsky was struck by the question, Is it possible that the Exodus of the Israelites described in the Bible appears in Egyptian records too? An enormous dislocation of people had taken place during calamitous events: Mount Sinai had erupted, a "pillar of cloud and fire" had appeared in the sky, the plague had broken out, and a land passage had appeared miraculously to part the Red Sea. If such a chronicle existed—and there were good reasons to think it did—Velikovsky might be able to link the histories of Israel and Egypt and synchronize their chronologies.

He found the answer in an obscure papyrus stored at the Leiden Museum in the Netherlands. Discovered in the early nineteenth century, translated in 1909 by the British Egyptologist Sir Alan Henderson Gardiner, and dated to the end of Egypt's Middle Kingdom, this document recorded the lamentations of the sage Ipuwer, who chronicled the escape of slaves from Egypt during a period of chaos and upheaval.

This discovery started Velikovsky on a path of research and writing that would keep him busy for the rest of his life. Nine years later he submitted two manuscripts for publication, *Ages in Chaos* and *Worlds in Collision.* The first book, dealing with chronology and culture, attempted to prove that Egypt's history was about five centuries shorter than historians thought. In effect, Velikovsky maintained that Egyptian history was no older than the Hebraic. The second text was a documented description of a possible cataclysm related to the Exodus. After more than a dozen publishers had rejected both works, the venerable house of Macmillan in New York made Velikovsky an offer.

Macmillan decided that *Worlds in Collision* would appear first, though it had been written second. There were good reasons for this reversal: in the absence of a catastrophe simultaneously witnessed in Egypt and in the Middle East, *Ages in Chaos* would have lacked any foundation.

Worlds in Collision described a scenario Velikovsky had imagined one afternoon in October 1940 while thinking about the Old Testament. He recalled how, fifty-two years after the Exodus, "the Lord cast down great stones" (Joshua 10:11), then "the Sun stood still in the midst of heaven, and hasted not to go down about a whole day" (Joshua 10:13).

The image of an unmoving Sun made a strong impression on Velikovsky, as it did on his readers. He asked himself if this story could have any connection with the events that had taken place half a century earlier. A survey of other written sources from around the globe convinced him that a cosmic cataclysm, in which the planet Venus played a crucial role, had really happened.

Velikovsky's scenario looked like a science fiction script. A huge comet, which had originated from Jupiter, revolved for centuries on a stretched ellipse about the Sun. Around 1500 BC it came near the Earth twice within a period of fifty-two years, halting our planet's spin each time. During the eighth and seventh centuries BC, the comet approached Mars, forcing its passage close to the Earth. Finally it cooled and became the planet Venus.

Not content with relying only on the Old Testament, the Talmud, and the Ipuwer Papyrus, Velikovsky cited in support of his theory an impressive range of literature, including texts and legends from Arabia, Babylonia, Persia, India, Tibet, Armenia, West Africa, Greece, Rome, Iceland, Finland, Siberia, China, Japan, Mexico, Peru, and the Pacific Islands. Some of these sources mentioned a doomsday when the Sun stood still in the morning or at noon, depending on the geographical position of the reporter; others wondered why a particular night had been longer than usual. They referred to earthquakes, floods, and clouds of fire—similar to the ones described in the Bible and the Ipuwer Papyrus—or to wars between gods.

To support his claims, Velikovsky corroborated the texts with physical evidence. For example, he maintained that the last Ice Age gripped Europe and North America but not Asia, because the angle of the Earth's axis was different before the encounter with the comet. In addition, the comet must have become Venus in the seventh century BC because previous documents mentioned only the planets Mercury, Mars, Jupiter, and Saturn. Velikovsky had a simple answer for every question, and any educated person could follow his reasoning. The pre-publication excerpts demonstrated his captivating and convincing prose.

Before the book appeared, *The New York Times*, *This*

Week, *Herald Tribune*, *Pathfinder*, *Collier's*, *Vogue*, *Reader's Digest*, *Newsweek*, and several other magazines and newspapers described *Worlds in Collision* in glowing terms. They spared no words in praising Velikovsky and his multidisciplinary work, which, the media said, reflected a deep knowledge of mathematics, physics, chemistry, classical literature, folklore, world history, and religion. His conclusions were deemed revolutionary in their opposition to the standard views held about chronology and the physical sciences.

The scholars reacted very differently: they crushed and mocked the ideas of the not-yet-published book. The promotion campaign for *Worlds in Collision*, the impression Larrabee's article made on the public, and the enormity of Velikovsky's claims irritated many of them. Within days, the chief editor of *Harper's* received more than three hundred letters condemning the publication of such trash. Other experts went public.

David Delo of the American Geological Institute said that the Russian doctor had disregarded the sound research of the Earth's crust made during the past century. Carl Kraeling, the director of the Oriental Institute at the University of Chicago, pointed out how unscientific Velikovsky's method was—first to accept a statement and then to look for evidence to support it—and how much ignorance he displayed about ancient Eastern literature. The archaeologist Nelson Glueck, who, before becoming president of the merged Hebrew Union College and Jewish Institute of Religion in Cincinnati, had uncovered more than one thousand sites in the Middle East, characterized Velikovsky's interpretation of the Bible as ridiculous.

The astronomers were similarly dismissive. Cecilia Payne-Gaposchkin of the Harvard Observatory derided the suggested motion of Venus and Mars, comparing it to "an extraordinary achievement in a very difficult type of marksmanship—four (or even five) hits in a couple of thousand years." Harlow Shapely, the director of the same institution, called the whole theory "rubbish and nonsense." He went so far as to send a letter to Macmillan, expressing his hope that *Worlds in Collision* would never be published.

Macmillan's president, George Brett, thereupon sent the proofs to three reviewers of his choice. Only one of them opposed publication, so Brett decided to continue with his plans. *Worlds in Collision* came out at the end of March 1950 and sold exceptionally well.

But soon Macmillan faced difficulties. More and more academics refused to meet its sales representatives or to adopt its textbooks. Aware of the risk, Brett made a painful decision. He arranged for Doubleday, which was not in the textbook business, to take over the deal. James Putnam, the Macmillan editor who had originally acquired the two manuscripts, left the house for undisclosed reasons.

Velikovsky made the best of these events. He complained in the media about the "science establishment" plot against him. In his view, this interest group, which claimed to seek the truth and to fight for freedom of expression, was doing exactly the opposite: scientists criticized his theory before they read it and they did their best to suppress its publication. That was a scandal, and every North American intellectual of the 1950s heard about it.

Public sympathy went with the oppressed. *Worlds in Collision* became a top bestseller, and many of its readers sided with Velikovsky. In the arguments that followed, the emotional component prevailed. Some scientists blundered during the debate, and the Russian doctor used every mistake to claim a new victory. John Q. Stewart, an astronomical physicist at Princeton University, for instance, derided the possibility that the rotation of the Earth around its axis had stopped because of the comet's gravitational interaction. In June 1951, he wrote: "The author perhaps does not fully appreciate what a sensitive indicator the oceans would be. Try it with a full dishpan in the back seat of your car." Velikovsky responded that he had. Accelerating at 2 miles per hour in 1 minute, and then decelerating at the same slow pace, he didn't spill any water. "I may be all wet," he added, "but the car stayed dry."

The controversy intensified, and Velikovsky won new fans. A few academics sided with him. Though highly respected public figures such as Isaac Asimov, Martin Gardner, and Carl Sagan joined the scientists' camp, they didn't leave the scene unruffled. Velikovsky and those who had closed ranks with him seemed to have the last word. Even when their arguments were shaky or incorrect, their better rhetoric left the impression they had won.

The circle of Velikovskian supporters continued to grow. They were intelligent, enthusiastic, and faithful. Apprenticeship was short—only as long as it took to read and understand *Worlds in Collision.* They believed in their cause and could contribute "scientific" ideas and articles within weeks or a few months of joining the club. No time or money or energy needed to be spent in graduate school, in the hunt for academic jobs, or in climbing the academic ranks. And success was guaranteed within the group.

Funds in support of Velikovsky mounted. Three new publications—*Pensée*, *Kronos*, and *SIS Review*, all dedicated to the studies of the mentor—soon appeared. If academic journals rarely accepted articles about catastrophism, these new intellectual forums invited manuscripts both in favour of and against the concept. That raised the status of the Russian doctor ever higher, and more people gained confidence in him.

Velikovsky started thinking of himself as the only person who understood the issues. No idea in his books was ever altered in subsequent printings; not a line was ever changed. To him, that was the absolute proof of his correctness, and he stated it with pride. Not surprisingly, the chemist Henry Bauer, who made a thorough analysis of this phenomenon, compared Velikovsky to the leader of an emerging cult.

If, in 1952, Velikovsky still showed a balanced spirit in the preface of *Ages in Chaos*—"I claim the right to fallibility in details and I eagerly welcome constructive criticism"—by 1974 he was unapologetic. In his presentation at the symposium "Velikovsky's Challenge to Science," held in San Francisco that year, he ended by saying, "None of my critics can erase the magnetosphere, nobody can stop the noises of Jupiter, nobody can cool off Venus, and nobody can change a single sentence in my books."

The magnetosphere, the noises of Jupiter, and the elevated temperature of Venus were predictions Velikovsky had made by assuming that a comet had nearly collided with the Earth. Although his forecasts turned out to be true, he never questioned whether the causes of these events might have been different. To most scientists, Velikovsky still appeared to be guessing. He was not establishing a solid theory that could explain anything more than isolated facts. Isaac Asimov compared this process to finding intelligible sentences in a random sequence of words.

In the 1970s some academics gave Velikovsky more credit, and he was invited to give a few lectures that became popular with students and with a small number of professors. Still, he never convinced the heavyweights. In 1955, in a *Scientific American* article, Einstein confessed to an interviewer about *Worlds in Collision*: "You know, it is not a bad book. No, it really isn't a bad book. The only trouble with it is, it is crazy."

LOOKING AT THE ARGUMENTS

A close look at Velikovsky's arguments reveals how speculative they were. Take, for example, his presentation of a story from the third millennium BC:

> "In the lifetime of Yao, the Sun did not set for ten full days and the entire land was flooded." An immense wave "that reached the sky" fell down on the land of China. "The water was well up on the high mountains, and the foothills could not be seen at all." (This recalls Psalm 104: "The waters stood above the mountains . . . they go up by the mountains" and Psalm 107: "The waves mount up to the heaven.") "Destructive in their overflow are the waters of the inundation," said the emperor. "In their vast extent they embrace the hills and overtop the great heights, threatening the heavens with their flood."

From these descriptions, Velikovsky concluded that China had witnessed a catastrophe similar to the one seen in Egypt and the Middle East: the Sun stood still and the ocean spilled over the land.

But the four Chinese quotes, taken from different sources, don't mention a wave. Viewed separately, they might describe a big flood. Furthermore, Velikovsky didn't question whether these sources could be trusted, although their veracity isn't obvious at all. How could somebody in the third, second, or first millennium BC decide that the Sun didn't set for ten full days, when the only way of measuring time was by the rise and fall of the sun? Or did Velikovsky assume that some kind of clock, no matter how primitive, existed at that time? If so, did he research this matter to provide an answer? Velikovsky presented no concrete evidence.

He also brushed aside the possibility of translation errors. Without checking the original language used in the text and weighing the meaning of words, phrases, and concepts, his conclusions lack any foundation.

Velikovsky often stated that Copernicus, Galileo, Darwin, Maxwell, Röntgen, and Einstein had initially been rebuffed. But he failed to say that thousands of other researchers had never succeeded because their ideas were unfit to survive. One such case is worth mentioning.

In 1883 Ignatius Donnelly published the book *Ragnarok: The Age of Fire and Gravel*, which bears striking similarities to *Worlds in Collision.* Like Velikovsky, Donnelly claimed that a close encounter with a comet had changed the angle of the Earth's axis and that earthquakes, huge winds, debris, and floods had occurred. He, too, backed up his statements with descriptions from all over the ancient world. And, like Velikovsky, he enjoyed great public success in his time. But nobody, except Velikovsky, remembered him in 1950. In *Worlds in Collision*, Donnelly appears in a footnote, with no mention of the similarities between the two men's theories.

An even more obvious example of how Velikovsky chose to "prove" his point is his mention of Herodotus. The Greek historian of the fifth century BC recalled a visit to Egypt in which he was told a story about how the Sun had changed its usual position, twice rising where it normally sets and twice setting where it normally rises. This incident seemed to relate to the Ipuwer Papyrus and the Exodus passage about the Earth's abnormal motion, followed by upheaval and catastrophes. But Herodotus' account ends with the sentence: "Egypt was unaffected by this: the harvests and the produce of the river were the same as usual, and there was no change in the incidence of disease or death." Velikovsky, however, omitted this detail.

RHETORIC VERSUS SCIENCE

I read *Worlds in Collision* more than half a century after its initial publication. Although I enjoyed Velikovsky's prose, I found his theory of planetary motion far from credible. I had to admit, though, that in spite of having objections to his proposed scenario, I couldn't rigorously refute it.

Other experts in celestial mechanics had faced the same challenge long before. As mentioned earlier, Cecilia Payne-Gaposchkin had dismissed Velikovsky's solution, and Harlow Shapely had called it "rubbish and nonsense." But these astronomers offered no proof for their statements either.

Velikovsky gave seemingly valid responses to his detractors: because planets move roughly in the same plane, if Venus had initially revolved on a stretched ellipse, it would have met the other planets sooner or later. He gave figures: the Earth has a 60 percent chance of going through the head or tail of a comet, assuming the tail were 100 million miles long. This reply made Payne-Gaposchkin's "marksmanship" comment seem naïve at best. Still, as I will show later in this chapter, Payne-Gaposchkin's claim can be mathematically proved and Velikovsky's rebuttal disproved.

A scientist who provided more detailed arguments against the catastrophic planetary scenario was John Q. Stewart of Princeton University. The June 1951 issue of *Harper's* magazine published three articles related to this problem: Velikovsky's "Answer to My Critics," Stewart's

"Disciplines in Collision," and Velikovsky's "Answer to Professor Stewart." One of Stewart's criticisms was that, if Mars had diverted Venus from an elongated ellipse, then both planets would have returned for many millennia close to the point of their initial encounter.

Velikovsky responded that Stewart had ignored the influence of magnetism and electricity, both of which change the laws of celestial mechanics. But this answer wasn't necessary. In Newton's theory of gravitation alone, without taking other possible forces into consideration, Stewart's statement was false. The orbits of three celestial bodies may become complicated if two of them pass too near to each other (see figure 1.1). Should another planet approach the Earth–Moon system, the motion could become chaotic because close encounters of three or more celestial bodies are as unpredictable as the outcome of a triple billiard ball collision. It could happen, for example, that the Earth and

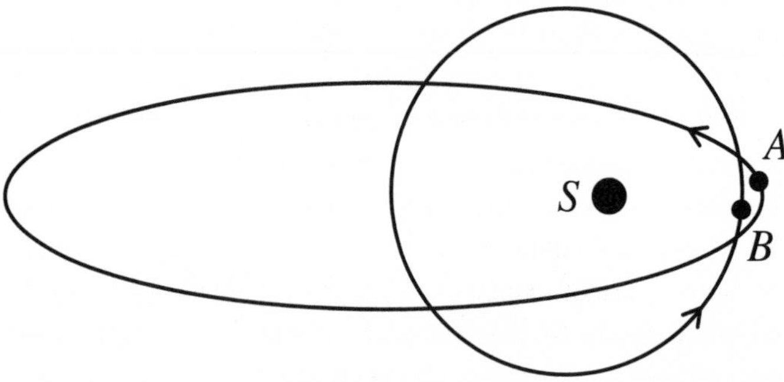

Figure 1.1—If planet A moves on an elongated orbit around the Sun and comes close to planet B, the two planets may eventually have very different orbits from their original ones, and neither of them will necessarily return close to the point of their initial encounter.

the other planet would start revolving around each other, while the Moon was expelled at high speed. But since in the 1950s only a handful of mathematicians knew that, neither Velikovsky nor anyone else noticed the slip.

Another weak link in Stewart's argument was related to the Titius–Bode Law, a formula that predicts the distance from the Sun to every planet except Uranus. Stewart wondered how the close encounter between Mars and Venus could take place without violating this principle. Velikovsky's answer was unassailable: the Titius–Bode Law is empirical and hasn't been proved to hold true on the basis of gravitation. Moreover, both Mars and Venus are only close to, but not at, the distance predicted by the formula.

In the summer of 1974 Robert Bass—an expert in celestial mechanics at Brigham Young University—published in *Pensée* what he claimed to be a proof of the Titius–Bode Law. He disagreed with Stewart, saying that this principle didn't contradict Velikovsky. If some massive celestial object is trapped in the solar system, the other planets readjust their positions according to the law. Unable to disprove Velikovsky's scenario, he cited the astronomer E.W. Brown, who in 1931, during his retiring speech as president of the American Astronomical Society, said he saw no reason why Mars, Earth, and Venus could not have nearly collided in the past.

But a careful check of Bass's proof revealed errors. He worked hard to fix them and thought he had. Still, no journal in celestial mechanics accepted his article, and he began to circulate the manuscript among experts. I happened to be on Bass's email list, but, like many others, I did not read his article because of its difficult style. One of those who did was Gordon Emslie, a physicist at the University of Alabama, who in the summer of 2003 found a fatal mistake.

Not so easily undone, Bass soon claimed to have corrected it. In the fall of the same year I emailed Emslie, asking him what he thought. He was still far from convinced by Bass. In his opinion, the proof couldn't be fixed.

Many researchers today, including the Princeton astronomer Scott Tremaine and the Nobel Prize laureate Steven Weinberg, remain sceptical that the Titius–Bode Law is a consequence of gravitation. Other experts, like Bass, are highly optimistic, hoping the law can be derived from gravitation. In any event, the Titius–Bode Law alone would neither prove nor disprove Velikovsky's theory. Therefore, some specialists attacked the planetary scenario in different ways.

Among them was Robert R. Newton of Johns Hopkins University, who, in the first volume of his 1979 publication *The Moon's Acceleration and Its Physical Origins*, referred to ancient astronomical observations. The dates for these records, however, are based on traditional chronology, which Velikovsky had assumed to be faulty. So, unknowingly, Newton used a circular argument. Finding a rigorous proof was far from easy.

PLANETARY MOTION

The secrets of planetary motion are hidden in the "*N*-body problem" of celestial mechanics, which seeks to determine the past and future positions of *N* celestial bodies on which the gravitational force acts. The case of two celestial bodies ($N = 2$), also called the "Kepler problem," is relatively easy to solve, and Sir Isaac Newton dealt with it in *Principia mathematica*. The orbit of one body relative to the other could be a circle, an ellipse, a parabola, a hyperbola, or a straight line (see figure 1.2). For three or more bodies, however, the problem has not been solved in general.

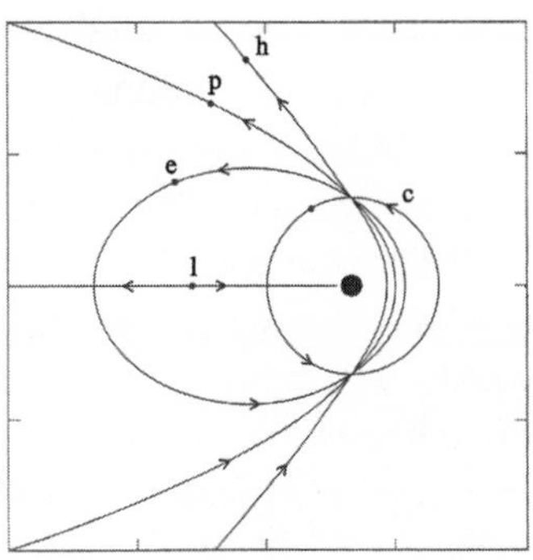

Figure 1.2—In the two-body problem of celestial mechanics, if one body is considered fixed, the other can move only on ellipses (e), circles (c), parabolas (p), branches of hyperbolas (h), or straight lines (l).

Planets move on almost circular ellipses around the Sun. Scaled to the size of this page, these ellipses would be hard to tell apart from circles without doing some measurements. But in reality, planetary orbits are more complicated. A perfect ellipse would appear only in a solar system with one star and one planet. When two or more planets exist, they affect each other's motion through gravitation.

Mercury, Venus, Earth, and Mars revolve on precessional ellipses (figure 1.3). A curve of this kind can be

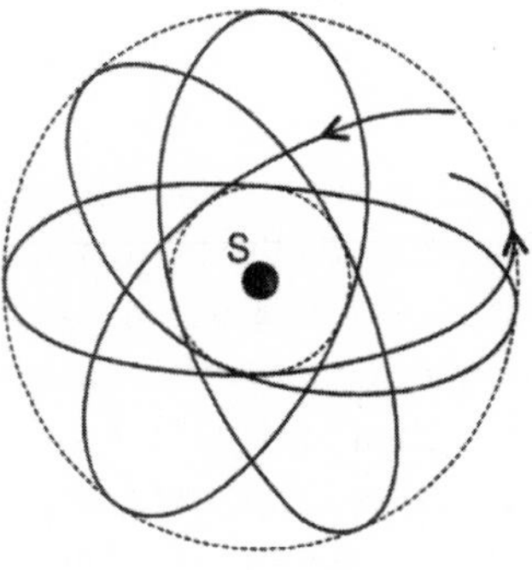

Figure 1.3—This planar curve is a precessional ellipse. The inner planets of our solar system (Mercury, Venus, Earth, Mars) move along such orbits. The deviation (measured as the speed at which the curve moves away from the original ellipse) is the largest for Mercury. Compared with the astronomical reality, the deviation represented in this drawing is greatly exaggerated.

sketched by moving the pencil as if drawing an ellipse while at the same time rotating the sheet of paper very slowly in its own plane. When looked at closely, these precessional ellipses are finely zigzagged curves. This zigzagging is caused by the gravitational pull of satellites and other planets.

Determining the exact planetary orbits is very complicated, and showing that a certain motion cannot occur is often impossible. No wonder that Velikovsky proved so hard to refute! Fortunately, I could evaluate the probability of Velikovsky's collision scenario, and I found it to be extremely small—as unlikely as making a pencil stand on its sharp end. I was able, then, to validate the correctness of Payne-Gaposchkin's marksmanship remark and the unlikelihood of Velikovsky's theory. (My argument is set out in a note on page 262.)

This result provided strong evidence against catastrophism and all its consequences. But, sometimes, improbable things do happen. So, to completely dispel the myth Velikovsky had nurtured, I had to prove that the quotations he took from ancient documents were misleading. To do that, I needed more than the unlikelihood of his planetary scenario. I had to show that it was impossible.

THE PROOF

The man who could help me disprove Velikovsky's theory of planetary collision was Jacques Laskar of the Institut de Mécanique Céleste in Paris. I first met him in 1998 during a conference in celestial mechanics at the Centre Paul Langevin in Aussois, a village in the French Alps. Although my expertise is hard to apply to the Velikovsky problem, Laskar's is ideal. He studies the practical aspects of the solar system and uses the latest analytic tools to provide accurate planetary positions for 20 million years both in the future and in the past.

This mapping is no easy feat. Powerful computers can do the calculations, but writing a good program for them is very difficult. One reason for this problem is the computation technique: although mathematicians have invented many numerical methods, all of them are approximations, and choosing a suitable one is a delicate issue. The round-off errors pose another challenge: since computers cannot write numbers with infinitely many decimal digits, each number is truncated. The methods used to determine planetary positions require billions of computations, and the round-off errors add up quickly, leading to spurious results. But the greatest barrier to accuracy is the chaotic character of the system: the motion of celestial bodies is as hard to predict as the weather. Nevertheless, Laskar has overcome the difficulties and made valid long-term predictions.

I sent him an email message asking if he had any evidence of planetary near collisions in the past few millennia. He responded the next day. His message began: "Is this related to Velikovsky's book?" *Monde en Collision* had just been republished in France, and Laskar had been asked to write a review for the October 2003 issue of *La Recherche.* My question couldn't have come at a better time.

According to Laskar, nothing of importance had happened in this area during the last 20 million years. Chaotic behaviour has been negligible throughout this time. There is a small probability that Mercury and Venus will come close to each other in 3.5 billion years, assuming the Sun doesn't significantly change its mass before then. But concerning the past 20 million years, research carried out by Laskar and many others confirms the impossibility of near collisions between Mars, Venus, and the Earth.

Still, there is one final issue to consider. Velikovsky's supporters continue to maintain that there may be other forces, such as magnetic and electromagnetic ones, that have not been taken into account and that could influence planetary motion. Their concern is legitimate: occasionally Newton's gravitation law fails.

Problems of this sort had already been noticed by Urbain Le Verrier, one of the co-discoverers of Neptune, in 1859, and they received more attention at the end of the nineteenth century, when the American mathematician Simon Newcomb compared thousands of observations with the computations obtained in celestial mechanics. Most of them agreed within 1 arc second (see my note on page 262). This comparison provided an excellent confirmation of the theory of planetary motion, but a few of the results exceeded the acceptable limits between prediction and the record.

The largest discrepancy was the one concerning Mercury. As mentioned earlier, Mercury's orbit is a precessional ellipse, whose closest point to the Sun rotates in the plane of motion by about 500 arc seconds per century. The difference between observations and calculations is roughly 43 arc seconds in 100 years. This gap is huge when compared with the 1 arc second precision obtained in most cases, and it means either that unknown forces act on Mercury or that Newton's theory must be revised.

Many experts tried to explain this phenomenon. Their attempts ranged from taking into account the influence of other forces, such as magnetism or solar wind, to changing the gravitation law. In 1898 a German schoolteacher, Paul Gerber, gave an explanation for Mercury's motion within the framework of classical mechanics. His deduction, however, needed more justification. Gerber died shortly thereafter, and, unfortunately, nobody has continued his work.

Albert Einstein succeeded in filling the gap between prediction and observation with his general theory of relativity, which threw a new light on the problem. Whereas Newton had envisioned gravity as a force acting in a three-dimensional world, Einstein viewed it as a geometric property of a universe with three spatial dimensions as well as a temporal one. Gerber's approach proved to be an inspired approximation of Einstein's results within the framework of classical mechanics.

While general relativity opened the way to research in cosmology and other branches of physics, it was of little help to celestial mechanics. In a relativistic context, even the two-body problem (see figure 1.2) becomes extremely difficult, and there is no hope of obtaining any significant results in the case of three or more bodies. Therefore the search for a suitable gravitational solution within a classical framework continued.

In the 1920s the Bulgarian physicist Georgi Manev suggested another model, which gave a reasonable explanation for the orbits of Mercury, Venus, Earth, and Mars by altering the law of gravitation, but his contemporaries overlooked his contribution. The history of science, however, knows many revivals, and Manev's resurrection was a happy one. I happened to come upon his articles in 1991 and found them mathematically challenging. This led me to publish several results related to collision and near-collision orbits which have been further developed by other colleagues.

My research in this direction showed that Newton's law fails to provide good approximations for planetary motions only when large celestial bodies come close to each other. Then corrections become necessary. It turned out, however, that this one anomaly changed nothing in Velikovsky's case.

Laskar had introduced in his model all the observed non-gravitational effects of the past few hundred years. To corroborate his computations, he compared them with the geological record. The shape of the Earth, which varies according to changes in the size of the ice caps and in its internal mass distribution, determines the inclination of the Earth's axis and the Earth's orbit around the Sun. Laskar's latest solution was used to calibrate the entire Neocene period—the last 23 million years. The astronomical and the geological results agreed.

All these outcomes show that the hypothesis at the foundation of Velikovsky's chronology is flawed, and, as a result, his catastrophic planetary scenario collapses. Though some of his followers continue to try to justify his dating of ancient history with new arguments, the experts are no longer listening. Velikovsky's name doesn't inspire credibility.

In spite of its failure, Velikovsky's attempt to formulate a system had some positive effects. It stimulated many people to think more about science and its relationship with the humanities and the social sciences. In the longer term, it provided a gold mine of research for those interested in how various domains of human activity interact and regard each other. It raised questions about the accountability of scientists and the media, about fame and publicity, about our naïveté and wish to believe in miracles, and it revealed aspects of the world of scientific research that were little known before. It showed once again what a powerful tool the written word has become and how it can seduce the uninitiated reader.

More personally, the Velikovsky episode was a humbling experience for me. I learned to be cautious and careful about the application of scientific theories to humanistic study, and especially to look at chronology problems from different points of view. Though I had found a scientific proof in this case, history is not mathematics. In any investigation of chronology, one has to rely on weaker arguments such as "beyond a reasonable doubt" and "a balance of probabilities."

In fact, my encounter with the work of professional chronologists reminded me of my first trip to China: the language was incomprehensible, the customs were alien, and I couldn't grasp much of the new environment. But I wanted to understand this arcane field of historical dating. A few weeks of reading made it clear to me that I would have to take a new approach. I must not look at Fomenko's work with a mathematician's eye, but had to regard it from the point of view of a historian who understands mathematics. Moreover, I would have to learn to think like a chronologist. The best way to acquire my new identity was to begin with Joseph Justus Scaliger, the man who had founded the science of historical dating.

CHAPTER 2

A New Science

Time is the proper dimension of history.

ELIAS JOSEPH BICKERMAN

Until the sixteenth century, few people thought of dating the events described in ancient texts. The study of chronology was in its infancy. It seemed to be waiting for someone to weave the many strands of historical timekeeping into a single narrative thread. There had been many attempts to do so, and the various accounts of the progress of time since the beginning of the world made a certain kind of sense to the cultures that created them. But the world's horizons were now expanding, and new ideas were being put forth to explain how all past events fit within a single stream of time. The new science of chronology was coming into being.

THE FATHER OF CHRONOLOGY

The man who would become the founder of the new science had little formal education. In 1552, at the age of twelve, Joseph Justus Scaliger (see figure 2.1) entered the elite Collège de Guyenne at Bordeaux, which he abandoned three years later after an outbreak of the plague in the area. Back home in Agen, northwest of Toulouse, he began to study with his father, Julius Caesar Scaliger, a highly respected and prolific scholar, who recognized his son's unusual intellectual abilities. "Every day he required from me a short declamation," Joseph later confessed in his autobiography. "This exercise, and the daily use of the pen, accustomed me to write in Latin."

Figure 2.1—Joseph Justus Scaliger (1540–1609), the founder of the science of historical chronology, as portrayed in a painting at the Senate Hall in Leiden, Holland.

With a strong background in classics, Scaliger taught himself thirteen languages—Hebrew and Arabic among them—becoming one of the most respected philologists of his time. But it was his contribution to chronology that earned him a research professorship at the newly established University of Leiden in Holland, where he worked from 1593 until his death in 1609.

Though his interest in historical dating began when he was twenty years old, he did not complete his first book on the subject until more than two decades later. In 1583, the year following the calendar reform of Pope Gregory XIII, he published *De emendatione temporum* (On the Correction of Chronology), which assigned dates to the most important ancient and medieval events. The revised edition of this book was followed in 1606 by his final work, *Thesaurus temporum* (Repertory of Dates), a collection and arrangement of all available ancient chronological sources.

Researchers of chronology had existed before Scaliger. Ephorus (*c.* 405–330 BC) appears to be the earliest. In his thirty-volume work of universal history, of which only fragments survive, he shows a keen interest in dating events. Sextus Julius Africanus (*c.* AD 160–240), often called the first true chronologist, was the father of biblical history. Though Theophilus of Antioch (*c.* AD 115–180) produced a brief chronology of the Bible about half a century before him, it is not known whether Julius used it.

In his work *Chronologia*, Julius attempted to put together Hebrew, Greek, Egyptian, and Persian sources. His methodology became a model for future chronologists. An example referring to Moses shows how he connected various pieces of information:

> If one computes backwards from the end of the captivity, there are 1,237 years. So, by analysis, the same period is found to be the first year of the Exodus of Israel under Moses from Egypt, as from the 55th Olympiad to Ogygus, who founded Eleusis. And from here we get a more notable beginning for Attic chronography.

The work of Eusebius, bishop of Caesarea (*c.* AD 260–341), became more influential than that of Julius Africanus. He lived during the time of Constantine the Great, who invited him to take part in the Council of Nicaea. Eusebius endorsed the Arian doctrine of a Christ inferior to God, a belief that brought him close to excommunication, but he won the favour of the Roman emperor by presenting him with new editions of his chronology books.

To make his point more clearly, Eusebius introduced tables that provided parallel dates for major events, organized by different cultural reference points. Thus, for example, Jesus was born 2,010 years after the birth of Abraham, or in the year of the 194th Greek Olympiad, which is the same as the forty-second year of Augustus's reign and the twenty-eighth year after the Roman subjugation of Egypt, or the deaths of Antony and Cleopatra. All of Eusebius's entries begin with the "Year of Abraham." The calculations are based on the life spans mentioned in the Bible; for instance, Adam lived 930 years, Noah 950, Abraham 175, and Moses 120. This information was taken seriously, as Jack Repcheck notes in his book *The Man Who Found Time*: "For Eusebius and all future chronologists [of the Bible], these explicit life spans were always the starting point."

A later chronologist of influence was Georgius Syncellus (end of the eighth and beginning of the ninth century), who, as a monk in Constantinople, held a position of authority under the patriarch Tarasios. He wrote a chronicle of universal history that preserves a rich collection of ancient sources, many of them otherwise unknown. Much later, the religious reformer Martin Luther (1483–1546) published a notable chronology book, *Supputatio annorum mundi* (Reckoning of the World's Years). And there were many others.

In spite of these contributions, chronology until Scaliger was only a gathering of disparate dates for practical or religious purposes, such as Easter calculations or the ordering of biblical stories. As Denys Hay, a professor of medieval history at the University of Edinburgh, noted in 1977: "In classical antiquity there was virtually no system of chronology available to historians."

Things didn't change much during the Middle Ages. Most chronologists aimed at predicting the date of Jesus' return to Earth and the end of the world, which the Bible puts at six millennia after Creation. But from Julius Africanus, who set the event in AD 500, to James Ussher (1581–1656), the archbishop of Armagh, who calculated that it would occur at the beginning of the twenty-first century, every chronologist shifted the date of Creation, placing the Apocalypse a few hundred years after his own lifetime.

Although no one before Scaliger had connected the world's Western and Eastern cultures in a global understanding of history, an attempt had been made. In 1568 Gerardus Mercator—the father of cartography and the inventor of the first flat projection for the globe—published a 450-page volume in which he calculated several historical dates from eclipses and astronomical observations. Unfortunately, the Inquisition declared the book heretical because it cited prohibited works from antiquity. The few printed copies of Mercator's book vanished into private collections and had no impact on further growth in the field.

Scaliger, therefore, had to start almost from scratch. He became interested in chronology while studying the linguistics of calendars. In 1568, prompted by a book written by the Roman grammarian Censorinus in the third century AD, Scaliger wrote: "I do not see how the month of April can derive its name from *aperio* [to open, to discover]. First of all, since the year initially had only ten months, they must have always wandered and had no fixed positions in the year . . . [In fact] *aprilis* comes from *aper*, which is boar."

Most old chronicles used dates that no one in Scaliger's time understood. Explaining, for example, what Athyr 20 in 476 Nabonassar meant was a difficult feat at the end of the sixteenth century. Scaliger had to decipher the structure of the Egyptian year, find out what month Athyr was, determine Nabonassar's era, and finally design an algorithm for converting the dates. Each calendar posed similar problems.

The early Chinese based time reckoning on the Moon's period; ancient Mayans used the Sun; and Jews and Muslims combined the two systems into lunisolar calendars. Since the year consists of a decimal number of lunar revolutions (about 12.4) and a decimal number of days (about 365.25), calendars had to be continually adjusted. Some cultures introduced leap years to keep the months in pace with the seasons, whereas others rounded off the numbers to simplify the calculations.

But deciphering calendars was not Scaliger's goal. He wanted to grasp the flow of philological ideas. Who had influenced whom? What ancient poet had borrowed from what writer? To answer such questions, he needed to know when they lived. And to know those dates, he had to arrange the main historical events in chronological order.

In the mid-1570s Scaliger asked a difficult question: What are the origins of the European peoples? He approached the problem through the genealogies of their leaders, information he could trace from documents. Soon he had connected European rulers to their ancestors in both the Near East and the Middle East and had found that the web of genealogical trees spread all over antiquity.

Everything he wanted to know about the origin of peoples and cultures led to the problem of global dating.

During the next few years, Scaliger's main scholarly interests shifted towards chronology and the astronomical aspects of dating. Astronomy played an important role for understanding the structure of and the relationship between various calendars. It also proved essential for determining when the total eclipses or the passage of the comets described in chronicles had taken place. An especially useful book for Scaliger was *Liber de epochis* (The Book of Eras), published in 1578 by Paulus Crusius, a professor of mathematics and history in Jena. The main contribution of the German professor had been his dating of several epochs mentioned in the *Almagest*, Ptolemy's famous astronomy treatise from the second century AD. Among the thirty-two eras Crusius determined were those of Nabonassar (747 BC), Seleucid (312 BC), Philip (324 BC), and Dionysius (285 BC), all of which occurred often in ancient texts.

At the end of May 1581, Scaliger wrote to a friend about his latest project, a book that dealt with the ancient and modern calendars of all nations and with the problems of calendar reform. When published in seven massive tomes, *De emendatione temporum* provided dates for the main historical events of humankind. This colossal work treats in detail the astronomical bases of more than fifty calendars. Scaliger's list of eras, the skeleton on which his theory was built, had seventy-eight entries, several of which disagreed with Crusius's tabulations.

Scaliger amended Crusius's biblical dates—for instance, the Creation (from 3963 to 3949 BC), the Flood (from 2307 to 2294 BC), and the Exodus (from 1511 to 1496 BC). He also calculated dates for other events, such as the fall of Troy (1181 BC), the era of Augustus (AD 43), the correction of the Julian calendar (AD 8), and the era of Constantine (AD 308). When agreeing with Crusius, as he did on the foundation of Rome (753 BC) or the Seleucid era (312 BC), Scaliger tried to improve the arguments of his predecessor. Both Scaliger and Crusius were rigorous in their conclusions, in that they differed from previous chronologists who had used fewer calculations and more speculation to fix the dates.

The length of Scaliger's arguments varied with each event. For example, his dating of the Marathon battle, which marks the Athenian victory at the opening of the Persian Wars, was straightforward. Scaliger cited the Greek historian Herodotus, according to whom the Persian king Darius I ruled for thirty-six years and died in the sixth year after his Marathon defeat. So the battle fell in 31 Darius, which, according to Eusebius, was the 286th Olympic year. Since he had previously fixed the first Olympiad to 776 BC, Scaliger set the Marathon battle in 491 BC. This date also matched a lunar eclipse mentioned in some of the documents, an event that astronomers had traced to April 25 of the same year. Invoking different reasons, historians today place the battle in September 490, but some mention 489 or 492.

Other arguments are long and difficult to follow for those who are unfamiliar with astronomy and ancient calendars. The founding of Rome (753 BC) and the First Council of Nicaea (AD 325), for instance, involve lunar cycles, divergent eras, and details of solar eclipses. But disagreements among various sources make some of these dates controversial among historians even today.

The public reaction to *De emendatione temporum* was mixed. Some scholars praised the book as a great achievement; others attacked it as superficial and filled with mistakes. Scaliger attempted to deal with the critics in the second edition of the book, but he failed to satisfy his detractors. On his side were his friends and several colleagues, including the respected chronologist Sethus Calvisius, who had studied and dated a few hundred eclipses. Among the foremost critics was the respected astronomer Johannes Kepler, whose vast knowledge of chronology and calendars would involve him in the debate on Jesus' date of birth. Kepler found Scaliger's presentation difficult and in disagreement with his main conclusions.

In the following years Scaliger continued to revise his proofs, adding argument and substance to his work, which appeared in 1606 as *Thesaurus temporum.* Though he made many improvements and analyzed in depth the writings of early chronologists, his initial list of eras remained basically unchanged.

His critics kept attacking him. Perhaps the strongest criticism concerned his refusal to accept the precession phenomenon, on which astronomical calculations depend. Scaliger insisted that all astronomers who affirmed its existence, including Copernicus, were wrong.

Precession occurs because the direction of the Earth's axis through the poles is not fixed but wobbles *very* slowly. This phenomenon resembles the motion of an unstable toy top, whose axis fails to stay perpendicular to the ground (see figure 2.2). Since the axis rotates once about every 26,000 years, spring each year arrives about twenty minutes earlier than it did the previous year—a fact that inspired the name of this phenomenon.

Over time, these minutes add up to advance spring's arrival by one day every seventy-one years. However, the imposition of calendrical corrections, which also address other astronomical problems, forces the equinox to fall on either March 20 or March 21. Any chronologist who neglected to take precession into account for, say, dating an

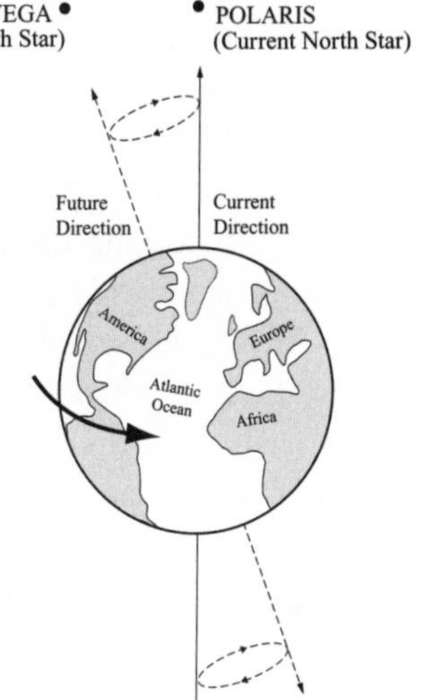

Figure 2.2—Day and night occur because the Earth rotates around its axis. In a first approximation, the direction of the axis is fixed with respect to the ecliptic (the plane in which the Earth revolves around the Sun). In reality, however, it moves very slowly and makes a complete rotation in approximately 26,000 years. The angle of the axis with the ecliptic remains constant.

ancient solar eclipse for which the place, the season, and the time of the day are known could be in error by several centuries.

Scaliger had two pieces of evidence against precession: first, the Egyptian claim that Sirius had been rising on the same day of the Julian year for more than a millennium and a half; and, second, the steady rising of Arcturus sixty days after the winter solstice. Had they been true, either of these arguments would have proved Scaliger's point. But they were wrong. Scaliger produced them by relying on literary texts, not on astronomical records.

He based his conclusion about Arcturus on a passage in *Works and Days* by the Greek poet Hesiod, whom tradition places in the eighth century BC. "When Zeus has finished sixty wintry days after the solstice," Hesiod wrote, "the star Arcturus leaves the holy stream of Ocean and first rises brilliant at dusk." But astronomical observation showed that the sixty-day estimate would hold true only for several decades, not several centuries.

It is interesting to note how, in this particular case, Scaliger had ignored a scientific achievement and let himself be deceived by a text whose credibility he could not verify. Velikovsky made the same mistake on a much larger scale. Time and again, this phenomenon occurs in the work of other chronologists, allowing bias to inhibit reasoning.

Refusing to acknowledge precession was not the only major trap for Scaliger. Mathematicians lost confidence in him after he claimed to have solved the ancient "trisection of the angle problem," which requires dividing an angle into three equal parts with a compass and an unmarked ruler. Nobody knew then that the problem was insoluble (a fact proved only in 1837 by Pierre Wantzel), but the mathematicians pointed at Scaliger's elementary mistakes. Adding insult to injury, they also found arithmetical errors in his calendrical calculations.

After 1606—in addition to the humiliating attacks on his work—Scaliger suffered a downslide in his personal life. His landlady sold the house in which he had spent the past decade, and he had difficulty finding another place to live. Finally he moved into a dwelling that was cold and drafty in the winter and leaked during the summer. His health deteriorated quickly, and less than two years later, on January 21, 1609, he died.

Despite its deficiencies, his chronology survived. As Princeton University's Anthony Grafton remarked in 2002: "The few modern historians who mentioned Scaliger described him as a brilliant innovator who created a discipline in the teeth of ferocious opposition." It was the new science on which history would rest.

PETAVIUS AND THE FIRST OPPONENTS

Scaliger's table of eras became the backbone of the modern world's understanding of ancient and medieval history. Any event linked to one of the entries in Scaliger's list could now be dated, thus filling in a new piece of the huge puzzle historians were trying to solve. But this edifice was fragile: if one basic date of Scaliger's proved to be wrong by a few centuries, his system would likely collapse.

The weakness of his structure was not due to any shortcoming on his part. In a way, this field of knowledge resembles geometry, where the modification of an axiom affects all the theorems resting on it. The amendment of the parallel's postulate (see my explanation on page 265) led in the nineteenth century to non-Euclidean geometries, which have played an important role in the birth of Einstein's theory of relativity. But while the physical sciences benefit from the fragility of such constructions, history suffers because of it.

This is why the next generation of chronologists felt the need to revise Scaliger's work. Among the new critics was Denys Pétau, a French Jesuit theologian and philologist born in Orléans in 1583, the year of publication of the *De emendatione temporum.* Pétau is better known under his Latinized name of Dionysius Petavius, to which he often added Aurelianensis in honour of his birthplace.

After finishing his theology studies in Nancy and being ordained in 1609, Petavius taught rhetoric in Reims, La Flèche, and Paris. During this time he read several classical and Christian writers and became interested in dating the events they described. In 1621 he was offered a professorship in positive theology at Clermont College in Paris, a position he kept for three decades, until shortly before his death.

In 1627 Petavius published his fundamental chronology work *De doctrina temporum* (On the Doctrine of Chronology), the first edition of which he dedicated to Cardinal Richelieu, the prime minister of France. Written as a polemic, the book corrected Scaliger and provided a chronological theory based on new or improved methods and techniques.

Petavius took the year AD 1 as the central date in history and established the usage of BC (before Christ) and AD (*anno Domini*, the year of the Lord). Though this practice was not uncommon among the monastic chroniclers of the late Middle Ages, it became widespread only after Petavius. He laid more weight on astronomical phenomena than his predecessor had, coming up with ingenious ways of relating dates with the motions of celestial bodies. Petavius used the combined cycles method extensively—lunar (19), solar (28), and indiction (15)—a system Scaliger had developed from the twelfth-century work of Roger of Hereford. The rationale for those cycles can be described succinctly.

Nineteen stands for the smallest number of full years the Moon takes to complete a full number of orbits (namely, 235) around the Earth. Twenty-eight is the least number of years after which the calendar repeats itself, with the dates matching the days of the week. Fifteen represents the Roman indiction, a taxation cycle established by the Emperor Constantine starting with AD January 1, 313. This period became standard in account keeping throughout the eastern Roman Empire.

The combined cycles method assigns to every date in history its corresponding Julian count, having as its base the Julian epoch (set by Scaliger to 4713 BC, mainly for mathematical reasons, and counting as 1). The year 753 BC, for example, has the Julian count 3961, because 4713 – 753 + 1 = 3961. Every Julian count up to 7980 has a unique triplet of numbers, resulting from the remainder obtained when dividing the Julian count by 19, 28, and 15.

The method works as follows: if certain documents associate a historical event to some year in the lunar cycle, another in the solar cycle, and yet another in the indiction cycle, then the year can be uniquely determined. The method has its limits: if one of the three elements is missing, the date cannot be ascertained.

Other ingenious techniques allowed Petavius to compute many new dates and to correct the ones Scaliger had based on less rigorous approaches. For these reasons, the two French scholars are now considered to be the co-founders of traditional chronology. But it was not always so. Like Scaliger, Petavius encountered opposition from his contemporaries.

Aside from facing disagreements regarding one date or another, his work was deemed fundamentally wrong because of the way he interpreted some documents. The French Jesuit Jean Hardouin, known for his scholarship on classical literature, proposed a radical critique. In 1685 he published a new edition of Pliny's *Natural History* in which he claimed that the majority of classical Greek and Roman texts had been forged during the Middle Ages by a group of Benedictine monks. Asked to elaborate, Hardouin said he would reveal the monks' reasons in a letter that should be read only after his death. But the executors of his estate were never able to locate that document.

As a result, most seventeenth-century criticisms of the work of Scaliger and Petavius were quickly forgotten, and the two chronologists received more recognition in the following decades. Still, it wasn't long before their conclusions were challenged again. In the 1720s Isaac Newton, who regarded chronology as one of the great issues of modern science, raised serious concerns about the dates set out by Scaliger and Petavius and proposed a system of his own. Newton's chronology attracted public attention and led to a bitter debate, which survived him by almost a century.

In 1856 and 1857 the German historian August Mommsen wrote several articles in which he attempted to revise Greek and Roman chronology. His more famous brother, Theodor, who would receive the Nobel Prize for Literature in 1902, criticized August in an 1858 book entitled *Die Römische Chronologie bis auf Caesar* (The Roman Chronology up to Caesar). In Theodor Mommsen's view, several significant dates remained unclear, while many estimates had to be changed slightly.

The application of such great minds meant that chronology grew to be a highly regarded research subject. Historians adopted and continued to develop the system of Scaliger and Petavius, but several results needed independent verification. Astronomy, which had improved its techniques considerably, was ready to offer some help.

A RUSSIAN POLYMATH

In the second part of the nineteenth century the Austrian astronomers Friedrich Ginzel and Theodor von Oppolzer sought to set traditional chronology on a firmer base. They investigated many of the astronomical events described in ancient and medieval chronicles and tried to improve the mathematical aspects of calendrical computation. Their results confirmed many traditional historical dates but cast doubt on others. Among the problematic landmarks were those of the Peloponnesian War between Athens and Sparta, whose beginning Crusius and Scaliger had assigned

Figure 2.3—Nicolai Aleksandrovich Morozov (1854–1946), the polymath who first claimed that historical chronology was wrong by about a thousand years.

to 431 BC, and the birthdate of Jesus Christ, which had been repeatedly challenged by chronologists.

The dating of the Peloponnesian War was particularly important because it relied on the occurrence of three eclipses with well-known relative dates. Descriptions of such phenomena are rare in history, and Scaliger considered 431 BC a very firm date for marking the start of the war. In the early twentieth century these issues came to the attention of Nicolai Aleksandrovich Morozov, a Russian polymath with a strong and rebellious personality (see figure 2.3). Before learning of the problems consuming chronology, he had a difficult life. But the troubled years of his youth were now behind him, and he seemed ready to embark on a long and risky adventure.

Morozov was born in 1854 in the town of Borok, near Yaroslavl, north of Moscow. His father was an aristocrat, but his mother had been a simple peasant. Because his parents had married before the civil authority and not in the Russian Orthodox Church, Nicolai was given his mother's last name.

At the age of twenty he joined the ill-fated revolutionary movement against the tsar. This opposition led to his imprisonment in 1881, first in Petropavlovsk and later in the infamous Schlüsselburg fortress on Lake Ladoga, east of St. Petersburg, where he spent more than twenty years. During this time, Morozov wrote poetry, memoirs, and essays, taught himself eleven languages, and became erudite in astronomy, history, physics, mathematics, chemistry, linguistics, and biology. Between his release in 1905 and the Russian Revolution of 1917, he focused on writing, science, and education.

In 1906, at the recommendation of the respected chemist Dimitri Mendeleev, Morozov was awarded an honorary doctorate for a work entitled *Periodic Systems at the Foundation of Matter*. This distinction marked the first official recognition of Morozov's long and intense creative activity, which ended only at his death in 1946. He published technical papers and books in chemistry, mathematics, astronomy, astrophysics, cosmology, biology, gravitation, relativity, atomic physics, history, geology, aeronautics, philology, and linguistics as well as his own literary creations, and he translated several novels into Russian, including *The Time Machine* by H.G. Wells.

The victory of the October Revolution changed Morozov's life. He was appointed director of the P.F. Lesgaft Institute for the Natural Sciences in St. Petersburg and was later elected a member of the Russian Academy. In 1923 Lenin ordered that Morozov's land be exempt from nationalization, to reward the aging scholar's contribution to the revolutionary movement. Morozov spent his last years on the estate, creating a branch of the Academy and a recuperation centre for its members. After his death, his birthplace was turned into a museum.

Morozov's first book about chronology, *Revelations in Storm and Thunder*, appeared in 1907. It dealt with several events described in the last book of the Bible. His dates differed from those of Scaliger by several centuries. In his second book, *Prophets*, published in 1914, he analyzed the chronology of biblical prophecies by using astronomical arguments. Again, his conclusions contradicted the traditional ones.

Morozov then initiated a systematic study of the chronological system founded by Scaliger and Petavius. The fruit of his research was the seven-volume treatise *Christ: The History of Human Culture from the Standpoint of the Natural Sciences*, published in Russia between 1924 and 1932. Morozov argued that the accepted chronology had been artificially inflated by the mistaken repetition of the same events in different epochs.

For airing this unconventional view, Morozov encountered serious obstacles during the publication of *Christ.* Until 1921, when he completed the first three volumes, all his submissions were rejected. In frustration, he wrote directly to Lenin, complaining that the bureaucratic system of the Soviet publishing houses blocked the spread of new ideas. Lenin asked his education minister, Anatoli Lunacharsky, to look into the matter. The minister's report confirmed the bureaucracy's decision to reject the manuscript.

But Morozov didn't give up. He persuaded Lunacharsky to attend a meeting where he presented his ideas. He must have been very persuasive, for the education minister became convinced of the power and novelty of the argument in *Christ.* In a letter to Lenin, Lunacharsky emphasized that Morozov's book was "no absurdity" and urged him to order the book's publication. Lenin consented.

In spite of that support, the printing of *Christ* was repeatedly postponed. Morozov asked the authorities to intervene again in 1923. But this time he approached Felix Dzerzhinsky, the commissar for internal affairs and head of the All-Russian Extraordinary Commission for Combating Counter-Revolution and Sabotage (better known as Cheka), which had carried out hundreds of thousands of executions. Nobody dared to oppose Dzerzhinsky, and the first volume of the treatise appeared in 1924.

After his powerful supporters died (Dzerzhinsky in 1926 and Lunacharsky in 1933) and the first seven volumes had appeared, Morozov was again banned from publication. This time he was blocked because his next instalment made a sharp revision of Russian history, a change that Josef Stalin—in power since 1924—disliked. Morozov could do nothing, and the last three volumes of his work never appeared in print.

The seven published volumes of *Christ* dealt mainly with the chronology of ancient Greece, Rome, Egypt, and China. Morozov showed that the traditional dates ascribed to astronomical records were wrong. He analyzed ancient horoscopes to date the sky configurations they encoded, employed statistics and probabilities to argue that the dynasties of certain rulers overlapped, and presented linguistic arguments in favour of his ideas.

Though he was not the first to oppose the theory of Scaliger and Petavius, nobody before Morozov had argued against the accepted chronology of China, which seemed to have grown independently of the European scheme and was apparently untouched by its problems. But Morozov disagreed. In his view, Chinese chronology had not developed until the seventeenth and eighteenth centuries AD, and then under the influence of Scaliger and Petavius.

Morozov argued that Chinese astronomy had never reached the high level of sophistication later attributed to it. The recorded astronomical observations were imprecise and unreliable, and only a few of them could be used for chronological purposes. In his opinion, the Chinese couldn't have invented the telescope long before the Europeans did, and the idea of an ancient and medieval Chinese civilization superior to its contemporary Western culture derived from an erroneous chronology.

Morozov considered that the mistake had been made in a misapprehension of the Saturn–Jupiter cycle. The early Chinese had used for time-reckoning the sixty-year period during which Jupiter and Saturn simultaneously complete the smallest number of full revolutions around the Sun (Jupiter, five; Saturn, two). The Chinese observations of these planets led Morozov to conclude that the starting point of the first recorded cycle was not the third millennium BC, as history books taught, but 1323 AD.

As evidence for this radical reordering of the millennia, Morozov mentioned a Chinese emperor traditionally considered to have lived between 2513 and 2436 BC. The use of the sixty-year cycle had begun during the reign of this ruler, whose astronomers recorded the alignment of all visible planets near the stars α and β of the constellation Pegasus. But the grouping of planets in a certain region of the sky is a very rare phenomenon, and no such configuration occurred in those years. The only viable alignment took place on AD February 9, 1315.

It is astonishing to see Morozov shortening Chinese history by almost four millennia. After all, Chinese documents exist that record daily events for hundreds of years. But other Western sources appear to support Morozov's view. For instance, the respected British historian Sir Herbert

Butterfield wrote in his book *The Origins of History*: "The cataclysms of Chinese history seem to have spared little of the historical writings of the pre-Confucian days [before 550 BC]; and from early times there seems to have been controversy about the genuineness or the textual accuracy of the things that did survive."

Still, Morozov was attacked in both the popular and the scientific press. Unlike the situation during the age of Scaliger and Petavius, the knowledge that had accumulated by the twentieth century had become too vast to be mastered by any individual, no matter how brilliant. Consequently, many experts found flaws in Morozov's arguments and ridiculed his conclusions. But he remained unmoved by his critics. Though admitting to the possibility of weaker claims due to the uncertainty of the information he found in some historical sources, he strongly believed in the core of his work. The principle of repetitions and their inflationary effect on historical chronology was an idea he never gave up.

TIME SHIFTS

Morozov had no students, and his work in chronology was in danger of being forgotten. But in the early 1970s the Russian mathematician Mikhail Postnikov revived Morozov's ideas through a lecture series given at the History Institute of the Soviet Academy of Sciences. Postnikov concluded that even if Morozov was not correct in all his claims, he was right in principle, and traditional chronology left much to be desired. Although historians were not impressed, a young Russian geometer named Anatoli Timofeevich Fomenko became interested in the subject. He obtained a copy of Morozov's *Christ* and read it in detail.

Fomenko soon connected a problem in celestial mechanics with Morozov's theory. Records of ancient and medieval eclipses showed that a certain component of the Moon's acceleration exhibited anomalies that could not be explained in terms of gravity. But if history were shown to be shorter than had been thought, the dates of eclipses might be wrong and the anomalies of the Moon's acceleration might only be apparent. Years later, when he gathered enough evidence for a new chronology, Fomenko thought he could solve the problem merely by shuffling dates around. He described the aftermath of his finding in a book he entitled *History: Fiction or Science*:

> I had to address several distinguished historians with this quandary, including the ones from our . . . Moscow State University. Their initial reaction was that of polite restraint. According to them, there was no point whatsoever in questioning the consensual chronology of ancient history since all the dates in question can be easily verified in any textbook on the subject and have been proved veracious a long time ago. The fact that the diagram of some parameter [the component of the Moon's acceleration] started to look natural after revised calculations based on some flimsy new chronology was hardly of any relevance [to them]. Moreover, it would perhaps be better for the mathematicians to occupy themselves with mathematics and leave history to historians. The same sentiment was expressed to me by [the famous Russian historian] L.N. Gumilyov. I refrained from arguing with him.

Another issue Fomenko studied was the dating of astronomical occurrences related to key historical events. He understood immediately that the Peloponnesian War was a critical landmark. He disagreed with Scaliger's date of 431 BC for the start of the war and confirmed that the date proposed by Morozov, 1133 AD, agreed with the documentary description of the eclipses. But Fomenko also identified a third possibility—1039 AD—and showed that no other solution existed for the time interval of history.

Other events he researched included the eclipses described by Livy and Plutarch, those seen during the life of Jesus, and the observation of the star of Bethlehem, which could have been a huge star explosion called a supernova. Like Morozov, Fomenko concluded that these events had happened closer to the present time by about a millennium.

Also connected to astronomy was the Gregorian reform of the Julian calendar, which Fomenko and his assistant Gleb Nosovski analyzed in detail. They challenged the accuracy of the ten-day correction made by Pope Gregory XIII in the sixteenth century. The reform relied on the date of the Nicaean Council, which Crusius had fixed to AD 322 and Scaliger to AD 325. But the Russian mathematicians calculated that the council had met some five and a half centuries later.

Together with Nosovski and mathematician Vladimir Kalashnikov, Fomenko also dealt with the problems raised by the *Almagest*, Ptolemy's famous second-century astronomical treatise. By examining the configuration of the star catalogue, the eclipses, and the records describing the obscuring of stars by planets, they dated the *Almagest* to the period between AD 600 and 1300, with a high probability for the ninth century.

Fomenko also continued Morozov's study of Egyptian zodiacs and horoscopes. Some of the paintings and reliefs that feature these configurations were not familiar to historians, whereas others had been under the scrutiny of Egyptologists for many decades. Fomenko and his colleagues placed some of these objects at least a millennium later than Egyptologists had.

Shifting ancient historical dates forward in time poses the problem of fitting the documented rulers in a span of fewer years. The long lists of kings and queens in various parts of the world seem to contradict a shorter chronology. Or is this contradiction only apparent? In addressing this problem, Isaac Newton identified two parallel kings who had been made consecutive, and Morozov indicated several monarchs who had been duplicated but under different names. Fomenko went further by pointing out fourteen pairs of overlapping dynasties. He saw this parallelism as an indicator that entire historical periods had been mistakenly created.

To support his dynastic method, Fomenko came up with a statistical study of the documents' language and the maps' geographical features. Then he applied his findings to specific texts and charts. The results agreed with his earlier conclusions.

One of his arguments concerned the chronology of the Bible. It had always been known that the First and Second Books of Samuel speak about the same events as the First and Second Books of Kings, and that the First and Second Books of the Chronicles also overlap. According to Fomenko, however, several other chapters refer to simultaneous events, and the present ordering of the books in the Bible is flawed.

Fomenko also brought many etymological arguments in to support his thesis, which he outlined in two volumes published in Russia in the 1990s. But he encountered harsh criticism from linguists, who think that both his premises and his conclusions are wrong.

In spite of this strong opposition from specialists in many disciplines, Fomenko claims to have a mountain of evidence in favour of a new chronology. He thinks that tradition has mistakenly shifted an "original chronicle" (by which he means the documents that describe the historical reality) back into the past three times over. In other words, an event that happened x years ago, where x is larger than 500, has been interpreted to have also taken place $x + 333$, $x + 1053$, and $x + 1778$ years before. Though the idea of these shifts had already occurred to Morozov, Fomenko was the first to attach specific numbers to the concept.

But is he right? Can his mathematics lead to such conclusions? To explain how science adds to the understanding of chronology, it is necessary to grasp the meaning of the basic scientific dating methods, including radiocarbon, dendrochronology, thermoluminescence, fission tracking, archeomagnetic dating, and paleography. The analysis of these techniques reveals useful information about historical chronology and the relationship between the defenders of tradition and the advocates of reform.

Indeed, Fomenko and his colleagues are not alone in their quest to rewrite the past. They belong to a larger circle, which includes some professional historians and many amateurs. Few of their books have been published, and even those that are in print have captured only a small audience. Some of these researchers, however, have had significant social and cultural impact—for instance, Immanuel Velikovsky. But most of them have enjoyed no media coverage, and their conclusions are unknown to the general public.

In the following chapters I will investigate many of the problems outlined above, by presenting the arguments of some reformers of chronology as well as the important reactions to their ideas. The man who laid the foundation for understanding the physical reality needs no detailed introduction. His name is Isaac Newton.

CHAPTER 3

Swan Song

Nature and Nature's laws lay hid in Night:
God said, "Let Newton be!" and all was light.
ALEXANDER POPE

Isaac Newton's work and personality have been with me since my high school years. After I decided to pursue research in celestial mechanics, the science that Newton brought into being in 1687 with the publication of his masterpiece, *Principia Mathematica*, he loomed even larger in my mind.

Some biographers, such as Richard Westfall in *Never at Rest*, present Newton as an aloof genius who had no interest in worldly pleasures. Others, such as David and Stephen Clark in *Newton's Tyranny*, focus on his attempts to rule the British scientific community of the early 1700s through intimidation. And there are some, including James Gleick in *Isaac Newton*, who are sympathetic to the man.

I knew that the great scientist had nurtured many interests, but until I turned to chronology I had overlooked his contribution to historical thought. My research on this aspect of his life and work revealed struggle and intrigue, disputes and debates that have ramifications even in our time. Here is the story of Newton's attempt to rewrite the past and of the controversy that followed.

ACADEMIC PIRACY

On March 20, 1727, Sir Isaac Newton died in Kensington at the age of eighty-four (see figure 3.1). He was the first scientist to be buried at Westminster Abbey, already the final resting place of monarchs and poets. Among the hundreds of manuscripts Newton left behind, one in particular drew the attention of John Conduitt, the executor of his estate: *The Chronology of Ancient Kingdoms Amended.* Conduitt knew how keenly Newton had prepared this manuscript for publication during the last months of his life, and he carried out Newton's wish. In 1728 the book appeared first in London, Edinburgh, and Dublin and then in Paris in a French translation.

Based on astronomical and calendrical evidence, *The Chronology* was an attempt to redate the histories of the

Figure 3.1—This medallion depicting Isaac Newton was issued in 1727, soon after his death.

ancient civilizations of Greece, Egypt, Assyria, Babylonia, and Persia. Newton's conclusions sharply contradicted those of Scaliger and Petavius, his main point being that the French chronologists had mistakenly created historical periods that never existed.

What prompted Newton to spend the last years of his life untangling the complex and controversial problems of ancient chronology? His achievements in science and mathematics had already granted him immortality. Wasn't he afraid that becoming involved in the history of antiquity would only damage his reputation?

Not at all. From his early youth, he had set his restless mind to investigate various phenomena that he felt were far from clear. A notebook he kept from 1664 to 1665, when he was twenty-two, shows his first attempts to comprehend such notions as attraction, gravity, motion, fire, and light. The initial thirty-seven entries eventually increased to seventy-three. Some were mere headings—he failed to elaborate on the subjects of stability, fluidity, and humidity—but his comments on motion and colour expanded to several pages.

Newton's interests were universal and developed in many directions, from physics and mathematics to theology and alchemy. He was much concerned with religious studies, which led him to question the history of some wars and the dating of certain events. But he made a coherent presentation of his chronology system only late in life. He had many reasons to postpone this project: he focused on mathematics, astronomy, and optics, wrote *Principia*'s first two editions, led a fierce priority debate on calculus with the German philosopher and mathematician Gottfried Leibniz, was appointed Lucasian Professor of Mathematics at Cambridge, ran for Parliament, and served as president of the Royal Society and master of the Mint. And he might never have written

The Chronology of Ancient Kingdoms Amended had it not been for an act of academic piracy.

The person at the centre of the scandal was Signor Abate Antonio Conti, better known as Abbé Conti, a Venetian nobleman noted in the intellectual and aristocratic circles of Europe as a tragedian, translator of Pope and Racine, dilettante poet, and amateur scientist. In the early 1720s—under conditions of privacy and discretion—Newton gave Conti permission to make a copy of a manuscript he had entrusted to Caroline of Anspach, the Princess of Wales. Entitled "A Short Chronicle from the first memory of things in Europe to the conquest of Persia by Alexander the Great," the manuscript contained an introduction and a list of ancient historical events from 1125 to 331 BC, but with very few details about how they had been dated.

Not only did the abbé tell just about everyone he met about the "Short Chronicle," but he also allowed some people to see it. Among those who read the manuscript were Étienne Souciet, a Jesuit famous for his research on ancient chronology, and Nicolas Fréret, a scholar at the Académie des Inscriptions et Belles-Lettres in Paris. While Souciet contented himself with privately informing Newton about his objections to the dates presented in the chronicle, Fréret translated the entire work into French and submitted it to his publisher, Guillaume Cavelier, who sent Newton a letter requesting permission for printing. The English scientist didn't reply. On March 20, 1725, a frustrated Cavelier appealed to Newton again:

> Sir, six months ago I had the honour of informing you that a copy of your chronology had fallen into my hands. I asked you to inform me whether you had any additions

> or corrections to make in it because of errors on the part of the translator. Since the savants await anything which comes from a man as talented as you, with great eagerness, Sir, I have the honour of writing you this second letter to ask you to inform me immediately if you have something to change in it. If I do not hear from you, I shall take your silence for consent and let it appear as it is and I shall give it to the public with Remarks.

In spite of its arrogant tone, the letter made Cavelier's intentions clear. Newton answered this message with anger, prohibiting publication. Nevertheless, on November 11, 1725, he received a complimentary copy of his work, newly translated into French. The decision to print it had been made long before his response arrived.

Newton was outraged. He immediately drafted an article, of which seven versions have survived, whose final form he submitted to the *Philosophical Transactions of the Royal Society.* He denounced the pirated translation of the "Short Chronicle" and emphasized that Fréret's remarks attempting to demolish his theory showed nothing but utter misunderstanding of his work. But Newton felt that more than an article was needed to prove his theory correct, and soon he started drafting *The Chronology of Ancient Kingdoms Amended.*

Once the "Short Chronicle" became public, Father Souciet felt absolved of any obligation to keep his opinion private. In 1726 he published five long essays that drew on astronomical, numismatic, and literary evidence to attack Newton's work. John Conduitt, the relative who would become the executor of Newton's estate, was so afraid to let the octogenarian scientist read those articles that he asked a friend to summarize them in a positive light. Newton, however, went to the original sources and studied them carefully. To Conduitt's surprise, Newton didn't get angry; in his view, Souciet was simply wrong: from the "Short Chronicle" the Jesuit father had no way of grasping his arguments, and the new book would clarify the misunderstanding. But although he completed his project, Newton didn't live to see it in print.

His treatise is a mixture of logic and technique similar in style to *Principia.* Like Scaliger and Petavius, Newton fixed a few dates and then linked other historical events to them. In those cases where the ancient documents gave the configuration of the sky, he used astronomical calculations to find out when these events occurred. Of the first three dates he fixed, the most important one referred to the Argonautic Expedition, named after the ship *Argo*, which sailed in search of the Golden Fleece.

In Newton's time, chronologists were under the influence of Euhemerism, a theory originating in the fourth century BC, according to which mythology stems from the deification of human beings and the elaboration of their lives. Newton therefore believed in the reality of Apollonios Rhodios's story *Argonautica*, written in 295 BC. Traditional chronology dated the voyage to 1200 BC or earlier, but Newton came up with the year 936 or 937 BC. In the "Short Chronicle," he noted in his tortured prose:

> [In the year] 939, the ship Argo is built after the pattern of the long ship in which Danaus came into Greece: and this was the first long ship built by the Greeks. Chiron, who was born in the Golden Age, forms the Constellations for the use of the Argonauts; and places the Solstitial and Equinoctial Points in the fifteenth degrees or middles of the Constellations of Cancer, Chelae, Capricorn, and Aries. Meton in the year of Nabonassar 316, observed the Summer Solstice in the

> eighth degree of Cancer, and therefore the Solstice had then gone back seven degrees. It goes back one degree in about seventy-two years, and seven degrees in about 504 years. Count these years back from the year of Nabonassar 316, and they will place the Argonautic expedition about 936 years before Christ.

This argument is hard to follow (and typical of Newton's cryptic language), but an astronomer can quickly decipher it. Newton invoked the precession of the equinoxes (see figure 2.2). Since every spring's arrival precedes that of the previous year by a few minutes, the time of the equinox identifies the year.

This idea was revolutionary. Although by the sixteenth century astronomy had become an important chronological tool, nobody before Newton had thought to link precession with fixing the year. Any text that indicated the Sun's position at the equinox relative to the stars could now be dated.

RECONSTRUCTING THE COLURES

To provide an illustration of the exact positions of planets and stars, astronomers have designed an imaginary celestial sphere, which—like the Earth—can be endowed with parallels and meridians (see figure 3.2). The main point in Newton's argument was the position of the colures, represented by the two circles of the celestial sphere that cross perpendicularly at the celestial poles—one passing through the equinoctial points and the other through the solstitial points.

An ancient account of the colures, described in relation to the fixed stars, had appeared in a second-century BC treatise by Hipparchus, who quoted the fourth-century BC observations of Eudoxus. Newton commented on them:

> For Hipparchus tells us that Eudoxus drew the Colure of the Solstices through the middle of the great Bear, and the middle of Cancer, and the neck of Hydrus, and the star between the Poop and Mast of Argo, and the Tayl of the South Fish, and through the middle of the Capricorn, and of Saggita, and through the neck and the right wing of the Swan, and the left hand of Cepheus; and that he drew the Equinoctial Colure, through the left hand of

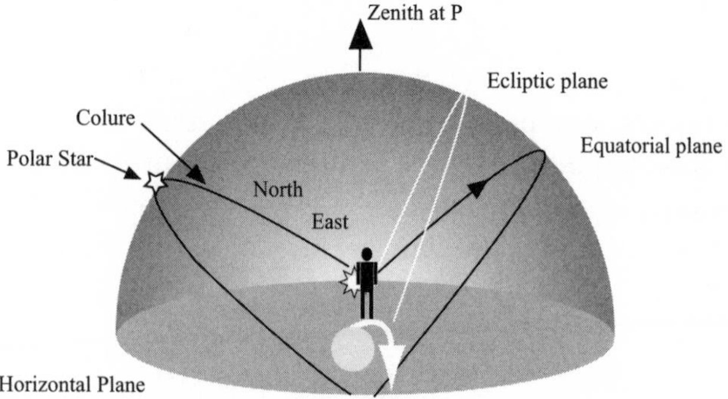

Figure 3.2—Astronomers assume that all stars and planets belong to an imaginary celestial sphere, which they endow with meridians and parallels similar to the ones imagined on the Earth. This sphere has an equator *and two poles. The Polar Star is very close to the* North Pole, *around which, due to the Earth's rotation around its axis, the celestial sphere apparently turns. The* equinoctial points *are the ones where the Sun is at the spring and fall equinoxes; these points lie at the intersection of the equator with the circle on which the Sun moves during the year. The* equinoctial colure *is the circle that passes through the North Pole and the two equinoctial points. The* solstitial colure *is in the plane of the page.*

> Arctophylax, and along the middle of his Body, and cross the middle of Chelae, and through the right hand and fore-knee of Cetus, and the back of Aries across, and through the head and right hand of Perseus.

Alas, the curves described above are not circles, and Newton spent much time trying to make them match the colures. The numerous drafts found among his unpublished manuscripts show how keen he was to understand this crucial point. In the end he estimated that, from the time of Eudoxus to his own time, the position of the spring equinox had changed by 36 degrees and 29 minutes.

Taking seventy-two years for the passage of each degree—because 26,000 (years) divided by 360 (degrees) is approximately 72 (see figure 2.2)—Newton concluded that Eudoxus had recorded an astronomical observation made in 939 BC (2,627 years before AD 1689—the year in which Newton was making his calculations). Given the inexact nature of Hipparchus' description, Newton never considered the dates based on it as very precise, conceding in his "Short Chronicle" that "there may be Errors of five or ten years, and sometimes twenty, and not much above."

But Newton didn't place Eudoxus in the tenth century BC. Citing Hipparchus, he reasoned that Eudoxus had merely reproduced the observations recorded when the Greeks had invented the celestial sphere. So, in Newton's mind, 939 BC was the year of the invention. Now he had to find the inventor.

The Greeks had divided the celestial sphere into constellations, which Aratus of Soli described in the third-century BC epic poem *Phaenomena.* Examining the names and the symbols of the constellations (see figure 3.3), Newton identified them with historical figures. Since these people lived

Figure 3.3—A 1690 map of the southern celestial sphere from Prodomus astronomiae *by Johannes Hevelius. Newton used this map to determine the position of Chiron's colure.*

before the Argonautic Expedition, he suspected that the sphere had been invented prior to the voyage to help the Argonauts in navigation. So he read everything that had been written on the subject, trying to find out who had conceived of the celestial sphere.

As his manuscripts show, Newton sought the inventor among the astronomers of Greek mythology: Achilles Tatius, Atlas, Endymion, Chiron, and Palamedes. He initially chose

Palamedes, but in the end settled for Chiron (see the quote on page 72), who had been Palamedes' master. In his eighties at that time, Chiron had an interest in helping the expedition because two of his grandchildren had been chosen to take part in it.

Finding the inventor of the celestial sphere closed Newton's argument. He could now reconstruct ancient history.

LANDMARKS

The next event Newton dated was the fall of Troy. Herodotus, the father of history, had mentioned in the fifth century BC that the time interval between the Argonautic Expedition and the end of the Trojan War was one generation, or roughly thirty-three years. Therefore Newton set the fall of Troy in 904 BC.

This conclusion allowed him to date the founding of Rome. In 19 BC the Roman poet Virgil wrote the story of Aeneas, who had escaped from Troy with his son Ascanius. After many adventures, the Trojans reached the west coast of Italy, where they settled. Ascanius founded the city of Alba Longa, in which Romulus—the legendary founder of Rome—and his twin brother, Remus, were born generations later. Estimating the time between Ascanius and Romulus, Newton fixed the event to 627 BC; Scaliger's date was 753 BC.

But historians have always debated this issue because of two traditions, one claiming that Aeneas, and the other that Romulus, had been the founder of Rome. In the second century BC, the Roman general, politician, and writer Marcus Porcius Cato combined the two stories into the generally accepted version, which Newton assumed in his calculations.

Like the fall of Troy and the founding of Rome, most of Newton's ancient dates follow from the Argonautic Expedition. They are based on documents, which Newton interpreted to the best of his ability. He had little confidence in historical sources other than the written word. John Conduitt recalled Newton saying about Lord Pembroke: "Let him have but a stone doll and he is satisfied. I can't imagine the utility of such studies: all their pursuit are below nature."

This belief was common during Newton's time. Not until another century had passed would archaeology compete with the legends.

SUPPORTERS AND OPPONENTS

When Newton died, the debate over historical chronology had just begun. His old friend Edmund Halley, the astronomer royal after whom a famous comet is named, was the first to defend him. But Halley was cautious about committing himself on the general validity of the amended chronology. Though in 1727 he endorsed Newton's calculations of the colure and the dating of the sky configuration, he refrained from commenting on whether those findings had anything to do with Chiron and the Argonautic Expedition.

This issue turned out to be the heart of the polemic. There were other murky issues, such as the average length of royal reigns. When the documents gave no information on the duration of a king's rule, Scaliger and Petavius had estimated it to be a generation (thirty-three years). Newton disagreed. He computed the average length of reigns from documents and came up with twenty years. But this issue paled when compared with the correct dating of the Argonautic Expedition. In his *Défense de la chronologie*, Nicolas Fréret criticized Newton's approach:

> Does one have to conclude from it [Eudoxus' text] that the sphere of Eudoxus was that of the first inventor of Greek astronomy? Is it not probable that this first sphere, very crude and faulty as the first essays of the human mind always are in sciences, had been later refashioned and that it was this Sphere corrected several centuries after the time of Chiron which Eudoxus used?

Though these views were published only in 1758, Fréret had begun his criticism three decades earlier. But if he feared before Newton's death that the aged scientist might have some surprise arguments to defend his position, after 1727 he voiced his objections without restraint. Still, things didn't work out as easily as he had thought. His early attempts to demolish Newton's chronology met with resistance.

It is interesting to note how polarized the issue had been from the beginning. In his earlier scientific debates, Newton had defended the priority of his results, not their validity. That shows how different science and mathematics are from history and chronology, in which truth and myth are so difficult to separate.

Among Newton's defenders was Andrew Reid, a popular-science writer and editor of a magazine in whose April 1728 issue a summary of the chronology appeared. Reid praised Newton for having "clearly explained the greatest mysteries of nature and obscurities of history." He thought the late English scientist "worthy to have statues of gold raised to his memory or rather . . . ranked among gods: for no mortal ever approached Divinity so near." This article enjoyed great success: it was republished in another London edition in 1732, one in Dublin fifty years later, and was also translated and published in France.

A strong attack against Newton came from William Whiston, who had been his student and later his successor as Lucasian Professor of Mathematics at Cambridge University from 1702 to 1710. Having spent many years around Newton, Whiston knew the man well. No one was better informed about the evolution of his mentor's chronology ideas, which he didn't find very enlightening.

In 1728 Whiston published a 120-page critique of Newton's book. Apart from putting forth the objections raised by other opponents (such as the dating of the Argonautic Expedition, the average duration of royal reigns, and the identification of Sesac and Sesostris), he also criticized Newton's claim that Homer and Hesiod had been almost contemporaneous. In regard to Eudoxus and the celestial sphere, Whiston accused Newton of gross misinterpretations. In his opinion, the traditional date of the voyage should have been raised by a century, not lowered by three.

The chair Whiston had occupied was very prestigious. Henry Lucas, a member of parliament for the university, had left instructions in his will for the purchase of land towards supporting this professorship, which was established in 1663. The first Lucasian Professor of Mathematics, Isaac Barrow, renounced the chair in Newton's favour. Newton occupied it from 1669 until 1702, when he decided to focus on his duties at the Mint. The present (seventeenth) holder is Stephen Hawking, famous for his research on black holes and as the author of *A Brief History of Time*, one of the best-selling popular-science books ever written.

It seems unlikely that Whiston would have obtained this position without Newton's support, so why would he viciously condemn his protector? Whiston's reasons for doing that downplay his critique. While holding the Lucasian Chair, he publicly denounced the Trinity doctrine and the Nicaean Creed, lending his support to Arius, who had claimed that God the Son was below God the Father. These views deprived Whiston of his professorship in 1710. Although sympathetic to Whiston's ideas, Newton didn't raise a finger to help him. Moreover, he voted against Whiston's election to the Royal Society.

Newton had always endorsed Arianism but was clever enough not to make his unorthodox religious beliefs public. His nomination first as warden and then as master of the Mint, as well as his attempt at a political career, showed him to be a far more shrewd politician than most of his biographers conveyed. Whiston, however, seemed to have learned little from him.

There is another reason to give Whiston's criticism less credibility. In 1732 John Conduitt received a letter from a Swiss mathematician living in London, Nicolas Fatio de Duillier, known for adding rubies to the mechanism of clocks. De Duillier compared the examinations of the celestial sphere made by Newton and by Whiston and indicated that he was inclined to agree with the former. He promised to check the truth with a new method, which—to his surprise—neither Newton nor Whiston had used. But no follow-up letter was found among Conduitt's effects after his death, so it is unclear what de Duillier could have proved.

The harshest criticism of Newton's amended chronology came from the French Jesuit scholar Jean Hardouin, the same man who had attacked the works of Scaliger and Petavius. In 1729 he published *Mémoires de Trévoux*, in which he mocked Newton's conclusion that Chiron had fixed the position of the colures on the celestial sphere: Chiron had been no astronomer but a physician known for curing plagues, and Newton's theory was nothing but "a frivolous system . . . imaginary and chimerical."

A later supporter of Hardouin's view was Abbé Antoine Banier of the Académie des Inscriptions et Belles-Lettres, who in his *Mythologie et les fables expliquées par l'histoire*, which appeared in 1740, ridiculed Newton's argument as circular. Though it's hard to believe that Newton could have made such mistakes, this is not the only time he was accused of blundering. In 1990 a mathematician claimed that Newton had mixed up the direct and the inverse theorems in a proof—an allegation that turned out to be unfounded. But chronology is not mathematics, and Abbé Banier speculated on the various interpretations of a text.

In spite of the many charges laid against Newton on both sides of the English Channel, another eminent Frenchman came to support him. This time it was the philosopher Voltaire, who in *Letters Concerning the English Nation*, published in 1733, devoted ample space to his chronology. If Voltaire was cautious in the first edition of his book, he fully endorsed Newton in subsequent printings. In 1758 he cast serious doubts on the criticism of Souciet and Fréret, who had based their attacks on the pirated version of the "Short Chronicle," which Newton had disowned.

François Marie Arouet de Voltaire was neither a scientist nor a historian but a literary giant who had the ability to convey difficult scientific ideas to the public. His admiration for Newton led Voltaire to publish a popular account of Newton's life and work. The book, which sold exceptionally well, made known the story that Newton's hypothesis of universal gravitation was prompted by the fall of an apple.

In the same year that Voltaire was casting doubts on the work of Souciet and Fréret, the latter's *Défense de la chronologie* appeared posthumously. Fréret had started writing the book in 1725 and was close to completing it in the summer of 1728, when he asked that a commission of the Académie des Inscriptions et Belles-Lettres examine his work. On December 17 he made his presentation in a closed session of the academy, which applauded his achievement. But then Fréret changed his mind about publication and continued to refine the arguments until his death in 1749. It took nine more years for the book to appear.

If his criticism passed almost unnoticed in 1728, it had a strong impact thirty years later. As Frank Manuel wrote in 1963, "Fréret disputed most of Newton's textual interpretation with a wealth of learning which made the scientist's classical knowledge look thin by contrast." Fréret contested Newton's estimate of twenty years for the average length of royal reigns by showing that the figure was based on the consideration of collateral as well as successive kings.

But his main point was again the Argonautic Expedition, and on this issue Fréret attacked Newton from several directions and with more data than anyone else. According to him, Thebes had been founded in 1594 BC, some five centuries before Newton's date, and the Phoenicians had made contact with the Greeks as early as 1884 BC. Obviously, Newton and Fréret had very different views about the early history of humankind.

THE NEXT GENERATION

Fréret's book ended an important period in the debate: the one in which Newton's detractors and supporters were his contemporaries. They had either known him in person or had corresponded with him. Those who followed belonged to the next generation, and they judged the work more than the man.

The first new voice was that of the twenty-one-year-old Edward Gibbon, who would go on to become one of the most respected experts on the history of Rome. In 1758 he read the amended chronology and, to fix its ideas in his mind, summarized it in writing and emphasized its strong points. By the end of this exercise, while agreeing with Newton's position, he refrained from accepting all its conclusions. Not even the works of Souciet and Fréret could convince Gibbon. About the English scientist, he wrote with admiration: "The name of Newton raises the image of a profound Genius, luminous and original. His System of Chronology would alone be sufficient to assure him immortality." About Fréret, he expressed regret: "Already full of esteem for this man of letters, I avidly devoured his response to the Newtonian chronology; but dare I say it?—it did not measure up to my expectations."

Although Gibbon became one of the giants in the field of ancient history, it is legitimate to ask whether he was mature and learned enough at this young age to understand the subtleties of chronology. He had converted to Catholicism six years earlier, influenced by his religious readings. Scandalized, his father sent him to Lausanne, Switzerland, where he spent five years in the company of a Calvinist minister whose persuasion led Gibbon to rejoin the Church of England. While in Lausanne he met Voltaire, who had a strong influence on him. It may be that Voltaire planted the seed of his admiration for Newton, which would explain the enthusiasm Gibbon showed for the English scientist.

Another of Newton's defenders was the mathematician William Emerson. In 1770 he published *A Short Comment on Sir I. Newton's Principia*, where he devoted a chapter to the voyage of the Argonauts. He criticized a Dr. Rutherford, Regius Professor of Divinity at Cambridge, who had attempted to demolish Newton's chronology solely on the basis of Chiron's colures. Emerson insisted that Newton had many other historical arguments to prove his point. He also rebuked Rutherford for claiming that the constellations had been mapped for religious purposes and not for navigation. As if to support Emerson, a new edition of Newton's amended chronology came out the same year.

In 1775 the classicist Robert Wood voiced a mixed opinion in *An Essay on the Original Genius and Writings of Homer.* One of the few who did not take a clear position for or against the amended chronology, Wood considered Homer to be the first poet of "barbaric" Greece. Thus Newton's shortening of Greek history suited his claim, and he praised Newton for this achievement but blamed him for giving Chiron too much credit. An astronomer who had both the science and the instruments to measure the colures did not fit easily with Wood's image of an uncivilized Greece.

A critique of how Newton had handled the Olympic years appeared in 1782 in the posthumous work of the Greek scholar Samuel Musgrave. As explained in chapter 2, Scaliger and Petavius used the dates of the Olympiads as landmarks of their chronology. Musgrave complained about Newton's claim that tradition had fabricated forty games. Again, it was a matter of interpretation over which Musgrave and Newton disagreed.

LOSING GROUND

Towards the end of the eighteenth century the polemic took a sudden turn. Between 1779 and 1785 Samuel Horsley published an edition of Sir Isaac Newton's complete works. The amended chronology appeared in the last volume, which included footnotes that pointed out what the editor thought to be errors. Horsley was a mathematician, an astronomer, and the secretary of the Royal Society, and he made no secret of the fact that he didn't believe in Newton's historical system. Therefore his editorial comments appeared to be an admission that Newton had been wrong. Not surprisingly, the debate faded away during the following years and few saw any point in flogging the dead horse of chronology.

More than four decades later, in 1827, an anonymous professor at Cambridge University published a work entitled *Essays on Chronology: Being a Vindication of the System of Sir Isaac Newton*, which was an isolated attempt to support the amended chronology. The author's wish to hide his identity shows how risky it had become to defend Newton a century after his death. No one approached the subject again until midway through the Second World War.

In 1942 the Soviet Academy printed a volume to celebrate the three hundredth anniversary of Newton's birth. It included essays on topics ranging from mathematics to philosophy, but one was of special interest. An article signed by S.Y. Lur'e examined Newton's amended chronology, claiming that the theory had failed because of erroneous premises and hypotheses. Newton was not to blame for the mistakes of "dull and lazy bureaucrats," by whom Lur'e meant the scribes who had copied (and altered) the original works of the ancients. The peculiarity of this statement is equalled only by the article's conclusion: Newton fell short in chronology because of his religious belief.

Frank Manuel took this idea further. In his book *Isaac Newton: Historian*, he offered the opinion that Newton's Judaic monotheism was his main motive for writing the amended chronology. Manuel seemed so convinced of this fact that he ended the book with the following words:

> To show that the Israelites rather than the heathen were the first founders of the humanity of the ancient world was [for Newton] the one historical end to which the long astronomical calculations and the reams of literary analysis were ultimately subservient. To relate himself to historical Judaism and primitive Christianity and to cut down the pagans and the Papists was the passion that animated his history. While the historical apparatus was neutral and the marginal annotations were accurate, sectarian religious commitment swept everything before it.

This conclusion is difficult to prove or to falsify. No doubt Newton was a pious man. He dedicated more time and energy to religion than to science, and even his scientific writing is marked by religious belief. But does this mean that he snubbed ethics to show that the polytheistic Greeks could not be the founders of the European civilization? If he had been so strongly driven to prove the priority of Judaism, why didn't he write his *Chronology* earlier in his life? Wasn't he drawn into this debate only after an act of academic piracy, which prompted him to defend his reputation? More likely, Newton's motive for taking on this subject was his thirst for understanding—an appetite he had nurtured all his life.

REVIVAL

Newton's attempt to rewrite history might have been forgotten were it not for some late twentieth-century reformers who reached similar conclusions. A British group led by Peter James, a writer and generalist in Mediterranean cultures, constructed the chronology of the ancient Near East, Middle East, and Europe starting from the nineteenth Egyptian dynasty, around 1200 BC, and from 700 BC in

Greece. In their 1991 book *Centuries of Darkness*, the dates the group obtained match Newton's. This team had solid credentials: I.J. Thorpe was an archaeologist specializing in European prehistory; Nikos Kokkinos, a historian of antiquity; Robert Morkot, an Egyptologist; and John Frankish, an archaeological expert of the East Mediterranean.

The starting point of the book was well motivated. Since the early nineteenth-century work of Jean-François Champollion, who by deciphering the hieroglyphs of the Rosetta Stone laid the foundations of Egyptology, the dating of Egypt has become fundamental for understanding antiquity. The eastern and western ancient cultures can be linked to Egypt, so their chronologies depend on it. By the end of the nineteenth century, archaeology had made progress in Greece, in the Near East, and especially in Egypt, whose chronology was regarded as scientifically sound. Consequently, historians viewed Egyptian chronology as a reliable base for dating antiquity. But this myth didn't last long. In 1892 the classical scholar Cecil Torr attacked a few basic Egyptian dates, triggering a four-year dispute with the British historian Sir Flinders Petrie, now known as the father of modern archaeology.

In 1896 Torr completed his arguments concerning the Egyptian site of Memphis and the Greek culture of Mycenae, where he allowed several overlaps between certain dynasties, thus shortening Egypt's chronology. The historian John Myres countered him in the *Classical Review*, and this response led to another two-year debate in the pages of that journal. In both disputes Torr had the last word, not for having won each argument on logic, but because his persistence wore his opponents out.

In addition to the Torr–Petrie and Torr–Myres debates, Peter James and his colleagues analyzed similar polemics on various issues, from the dating of the earliest Roman remains to those of pottery and inscriptions belonging to other civilizations. Looking at those disputes in the light of the most recent archaeological discoveries, they concluded that, although both sides were correct to some extent, their overall framework remained shaky.

According to *Centuries of Darkness*, the chronology of Egypt must be shortened by 250 years. That would bring the Trojan War to the mid-tenth century BC, in agreement with Newton's estimate for the fall of Troy (904 BC), and, as a result, the other dates of the amended chronology fall into place. So, by taking a different route, James and his colleagues appeared to have proved Newton correct. But soon after its publication, *Centuries of Darkness* also came under attack.

The most vehement critic was Kenneth Kitchen, a history professor at the University of Liverpool and an expert in Egypt's Third Intermediate Period, which spans the interval 1100 to 650 BC, from the twenty-first to the twenty-fifth dynasty. To historians, Kitchen's 1973 book marks the chronological foundation of that particular time and place. James and his colleagues, however, had based their theory on partial overlaps between the twentieth, twenty-first, twenty-second, and twenty-fifth dynasties, thus contradicting Kitchen's chronological system.

In a review published on May 17, 1991, in *The Times Literary Supplement*, Kitchen called the authors "young graduates" and "sons of Velikovsky," criticizing their theory in the harshest terms and predicting that it would have the same fate as that of their mentor. One of his points was the controversial list of Egyptian pharaohs created by Manetho, an Egyptian priest from around 300 BC. Among other things, Kitchen blamed the authors for their "irrational hatred" of Manetho.

Two weeks later James responded with a letter to the editor: "Sir, The tone used by K.A. Kitchen . . . shows that we have touched a raw nerve. Our book highlights a mass of archaeological and literary evidence, ranging from Spain to Iran, showing that Egyptian chronology must be seriously in error." James defended his views with good arguments, challenging Kitchen to explain why the twenty-first and twenty-second dynasties were successive rather than overlapping. He also quoted Kitchen as saying that Manetho's list could not be trusted.

On June 21 Kitchen responded with new evidence, standing by his initial review. He also brushed off James's critique of Manetho, complaining that his views had been taken out of context. Two weeks later James published another letter, remarking that Kitchen had not addressed the issue of the dynasties' successiveness, thus indirectly confirming his lack of proof for it. James remarked that although Kitchen's work on the Third Intermediate Period was fundamental for further studies, it was not as sound as its author claimed. The discussion seems to have ended here, leaving James the last word.

Other critics took a more balanced approach. In *KMT: A Modern Journal of Ancient Egypt*, Aidan Dodson, an archaeologist at the University of Bristol, although disagreeing with the 250-year figure, said that Kitchen's scheme was also wrong. In his view, between a quarter and a half century should be excised from the chronology of Egypt, depending on key events in Mesopotamia which lacked firm dates.

But Egyptologists seem far from agreeing on where to fix the pillars of their field. In 1977 Johannes Lehmann, a German theologian, philosopher, and television personality, noted: "In the course of a single century's research, the earliest date in Egyptian history—that of Egypt's unification under King Menes—has plummeted from 5876 to 2900 BC and not even the latter year has been established beyond doubt. Do we, in fact, have any firm dates at all?"

No wonder there is so much difference of opinion in the field. Sometimes even people who strive towards a common goal change their minds and pursue new directions. That occurred in James's group, which lost a key member before *Centuries of Darkness* appeared.

DISAGREEMENTS ABOUT EGYPT

His name was David Rohl, and in 1989 he decided to part with James and the other colleagues because, in his view, Egypt's history was shorter than they thought. When, as a nine-year-old, he had first visited Egypt, journeying up the Nile from Cairo to the temples of Abu Simbel in Nubia, Rohl fell in love with the ancient world of the pharaohs and dreamed about becoming an archaeologist. But his first career was in music. He founded a band, which released several albums, and made a living as a rock musician, producer, and recording engineer. Still, that childhood trip up the Nile haunted him until the desire grew obsessive, and one day he decided to change his life.

Rohl returned to school and studied Egyptology. While working on his PhD dissertation, he excavated at archaeological sites in Syria and Egypt. During this time he joined James's team, only to realize a few years later that he didn't fully agree with his colleagues. But this initial collaboration helped him find his own way.

His first book, *A Test of Time: The Bible from Myth to History*, appeared in 1995 in Britain and was followed by the three-part TV series *Pharaohs and Kings.* In 1998 he published a second volume, *Legend: The Genesis of Civilisation*, which led to the documentary movies *In Search of Eden* and *The Egyptian Genesis.*

Like James and his collaborators, Rohl began with the four main pillars that support the chronology of Egypt:

- 664 BC, the year the Assyrians sacked Thebes;
- 925 BC, when the biblical King Shishak, identified as Shoshenk I of the twenty-second dynasty, plundered the Temple of Solomon;
- 1279 BC, when, according to a document known as the Leiden Papyrus, Rameses II ascended the throne; and
- 1517 BC, mentioned in a medicine text called the Ebers Papyrus as the ninth year of Amenhotep I.

In *A Test of Time*, Rohl argued that only the first landmark was sound. One among many reasons for doubting the others was a rock-carved inscription unearthed at Wadi Hammamat, in the eastern Egyptian desert. The text—recorded during the times of Darius I, whose era is fixed to 496 BC—listed twenty-two generations of architects and connected the first generation to the early reign of Rameses II. Assuming (as Newton did) an average of twenty years for each ruler, Rohl concluded that a more plausible dating of Rameses' accession was 936 BC, about three and a half centuries later than tradition claimed.

But a single inscription is not enough to overturn history. It may have been that some generations of architects were missing from the list or that the text was misleading. So Rohl went through a long analysis of biblical accounts of Egyptian events to prove the overlap of the times in which the rulers of the twenty-first and twenty-second dynasties lived. This overlap favoured his shorter chronology of Egypt.

Though Rohl's books and movies created quite a stir in the British media, the experts' reviews were mixed. A couple of them gave Rohl high marks, others showed more reservation than enthusiasm, and a few dismissed the entire theory as nonsense. But the most vocal critic was the same Kenneth Kitchen who had attacked *Centuries of Darkness* in 1991.

Kitchen's first objection to Rohl's findings was that the founder of the twenty-second dynasty, Shoshenk I, had dedicated a statue to Psusennes II, the last ruler of the twenty-first dynasty. It proved that the two periods could not overlap. But Rohl had a good answer. There had been two kings named Psusennes, and it wasn't clear which one the inscription referred to. Traditional chronology had assumed without evidence that it was Psusennes II. The equally plausible assumption of Psusennes I would confirm his theory.

Kitchen invoked another inscription, this one on a statue in the British Museum, according to which Osorkon I, the son of Shoshenk I, had married the daughter of Psusennes II. This showed that no dynastic overlap was possible. Rohl responded with a similar type of argument as before. By identifying the name Osorkon with King Osorkon I, Kitchen had assumed the continuity of the two dynasties. The choice of Osorkon II, however, gave Rohl a new confirmation of the overlap.

Kitchen also mentioned two sequences of high priests, one in Thebes and the other in Memphis. Each straddled the border of the twenty-first and twenty-second dynasties, ruling out any overlap between them. Rohl answered that Kitchen had assumed conventional chronology to be correct when referring to a text that gave the years of a particular reign without specifying rulers. This argument was circular, so it didn't stand.

Kitchen's final objection relied on an inscription describing the annual flooding of the Nile during the reign of King Merenpath of the nineteenth dynasty. He drew on a technical detail to fix the period and show that the time span following it could not exist if the twenty-first and twenty-second dynasties were assumed to be parallel.

But Rohl was at home here. To reach his conclusion, Kitchen had cited an Egyptologist who had read only a copy of the text, without ever seeing the original. Rohl, however, had visited the archaeological site, deciphered the inscription, and interpreted it. He not only pointed out the errors of the published version that Kitchen had relied on but also explained why the inscription confirmed his own theory.

Kitchen was far from convinced. After the broadcast of Rohl's first TV series, he sent a letter to several Egyptologists in which he described the show as 98 percent rubbish. In his 1973 book, which saw a second edition in 1986, he worked out a detailed chronological map of the period, connecting thousands of details and building a foundation for future studies. Rohl's work threatened its very core, as *Centuries of Darkness* had done a few years before.

So, who is right—James, Rohl, or Kitchen? They each appear to have valid arguments, which makes arriving at a conclusion far from easy. Kitchen is an established expert, but James and Rohl possess real expertise too. It seemed wise for me to take some distance from all of them and to move away from the murky waters of the humanities. Perhaps, I thought, the scientific objectivity of Anatoli Fomenko would be more enlightening.

PART TWO

Fomenko's Battle Against Tradition

CHAPTER 4

Historical Eclipses

Chronology is nothing but the computation of celestial motions.

SETHUS CALVISIUS

Anatoli Fomenko has both the inquisitiveness and the capacity to comprehend many areas of culture. Proof of these attributes lies in his vast and complex body of work.

But even as a mathematician, Fomenko has been controversial. Quite a few distinguished colleagues, like my friend Tudor Ratiu, think highly of him. Others are less convinced. Most of them, however, have formed an opinion not by reading his work but by listening to what more influential researchers say. The attitude towards Fomenko is deeply polarized.

"You'd like him," Tudor once told me. "He's a fine and sensitive man." Tudor's remark made me cautious. If I wanted to remain objective, the last thing I needed was to involve my feelings. I decided not to meet Fomenko before completing this book. I had to judge the work, not the man.

The critical attitude towards Fomenko has nothing to do with the correctness of his theorems. It is about their significance: Are they deep and comprehensive? Do they have crucial consequences? Should we care about them? These aspects of inquiry leave the realm of mathematics. They approach the world of art and fashion, in which trends, personalities, taste, and biases play an important role.

Becoming intimately acquainted with Fomenko's mathematical work would have involved an investment of my time and energy that I couldn't afford, so I decided to focus on understanding his contributions to historical dating. Two aspects interested me. First, did Fomenko have a case against traditional chronology? Second, was his new dating system correct? Though not unimportant, the second issue mattered less, assuming that the answer to the first question was yes. If historians had not been able to clarify the problem in four centuries, I could overlook his failure to solve it in three decades.

Reading Fomenko's work straight through would have been a demanding task, so I decided to focus on one aspect at a time. And since astronomical results are among the most reliable data to be found in the field of chronology, I began with them. They reveal Fomenko at his best.

THE PELOPONNESIAN WAR

A landmark document for the classical age of ancient Greece is Thucydides' *History of the Peloponnesian War.* Very little is known about its author; the few existing bits of information come from what he disclosed in his book. He was an Athenian of Thracian royal origin, nearly thirty years old when the war began. He suffered and recovered from the plague at the beginning of the conflict and died in his late sixties or early seventies, without finishing his work. In spite of the mystery that veils his life, historians trust his word, and his book provides much of the knowledge acquired about that period.

The Peloponnesian War between the Greek city-states of Athens and Sparta lasted for twenty-seven years. Its main cause was the enmity between the conservative and militaristic Spartans and the democratic and innovative Athenians, and the clash of their philosophies and interests. The diplomacy of Pericles, Athens' leader, could only delay the armed conflict. Periods of intense fighting alternated with relative calm. Eventually Sparta won, but the length of the strife and the decades of animosity that followed weakened both powers, making them easy prey to the Macedonian invaders.

Thucydides chronicled the first twenty years of the war, abruptly ending his story in mid-sentence. In spite of tangled, occasionally opaque prose, he comes across as an eyewitness who is trying to describe things as accurately as possible.

But when did this war happen? The answer accepted today, and on which much of Greece's ancient chronology is based, came in 1578 from Paulus Crusius, who calculated that the conflict started in 431 BC. As is described in chapter 2, he based his dating on two solar eclipses and a lunar one, all of them described in Thucydides' book. The first event took place soon after the outbreak of war: "The same summer, at the beginning of a new lunar month, the only time, it seems, at which it is possible, the Sun was eclipsed after midday: it took the form of a crescent, then some stars became visible, and it turned full again."

The second eclipse occurred seven years later: "In the first days of the next summer a partial eclipse of the Sun took place at new Moon, and in the early part of the same month an earthquake." Then eleven more years passed until the third event: "All was ready, and they [the Athenians] were on the point of sailing away [from Syracuse], when the Moon, which happened then to be at the full, was eclipsed."

To determine how reliable these reports are, historians have questioned whether Thucydides witnessed the eclipses himself. Most likely he did, though he failed to mention it explicitly. Weather conditions permitting, these rare astronomical events are observed by millions of people over large areas of the Earth, and it would be unusual if Thucydides missed them in a place like Greece, where summers are long, dry, and clear.

In *De emendatione temporum*, Scaliger provided dates for the three eclipses—the days of August 3, 431; March 21, 424; and August 27, 413—indicating that the conflict began in 431 and ended in 404 BC. For more than three centuries no one challenged these findings. Even Isaac Newton supported the methodology: "These things are so well determined by eclipses and Olympic games and other records of good credit and so far agreed upon by chronologists, that I do not think it material to entertain any dispute about them."

But by the late 1800s, when astronomers had improved their techniques and computed all the eclipses starting from the twelfth century BC, several experts noted that the 431 eclipse of the Sun had been partial in the Peloponnese. This information posed a problem because stars are unlikely to show unless the solar disk is completely obscured.

More research followed, and some astronomers redid the calculations, taking into account the possible perturbations in the Earth's and the Moon's motions. But the conclusion was the same: an observer in Athens saw at least 9 percent of the Sun during the maximum phase. Perhaps the mentioning of stars was an exaggeration?

In his *Life of Pericles*, the Roman biographer Plutarch told how the darkness caused by an eclipse at the beginning of the Peloponnesian War had frightened the Athenians and how Pericles had used his cloak to explain the phenomenon and dispel the alarm. The Roman orator Cicero reported the same eclipse in his book *De republica.* But both accounts were written centuries after the fact, so the only surviving eyewitness description is that of Thucydides.

Another possibility is that, instead of seeing stars, he saw planets and didn't know the difference. Astronomers have investigated this possibility, but their findings have buried any hope that it might be true. Though Venus may have been visible at the time, Mercury appeared very faint, and Mars stood only 3 degrees of arc above the line of the horizon, a position where it was unlikely to be seen. Jupiter and Saturn were below that line, as was the star Sirius, which can be brighter than a planet. That left Venus alone and failed to explain Thucydides' report.

The latest word on the 431 BC eclipse has come from two British researchers, Francis Richard Stephenson and Louay Fatoohi, from the University of Durham. As experts in a branch of astronomy that deals with understanding and explaining ancient observations, they revised the old computations and, in 2001, published their results.

The reason for re-examining the problem was a phenomenon that hadn't been considered a century before. A modern interpretation of historical eclipses and, more recently, laser measurements have shown that the distance between the Earth and the Moon increases by a few centimetres a year. This change is due to a slowdown of the Earth's rotation around its axis at a rate that lengthens the day by about

2.3 milliseconds per century. To obtain accurate astronomical results, this aspect must be taken into consideration.

According to Stephenson and Fatoohi, the maximum phase of the 431 BC event showed a 12 percent crescent in Athens (see figure 4.1). In fact, the eclipse was nowhere total but only 98 percent annular (ring-like) at best. The researchers' calculations positioned the band of annularity as shown in figure 4.2. Still, this adjustment doesn't explain the appearance of stars, so this disparity continued to cast serious doubt on the contention that Thucydides' first eclipse took place in 431 BC.

Figure 4.1—The 431 BC eclipse of the Sun as seen in Athens at its maximum phase of 88 percent, showing a 12 percent crescent.

Figure 4.2—The (dark) band of annularity for the 431 BC eclipse, where 98 percent of the Sun was obscured. An observer located at a point on the dotted line saw a crescent showing 10 percent of the solar disk. In Athens 12 percent of the Sun was visible (see figure 4.1).

A recent event seemed to undermine the reliability of these results: the huge Asian earthquake that triggered the deadly tsunami on Boxing Day 2004. According to Richard Gross, a geophysicist with NASA's Jet Propulsion Laboratory in California, a shift of mass towards the Earth's centre during the quake caused the planet to move one-millionth of a second faster and tilted its axis at the poles by 2.5 centimetres.

The change in rotation was negligible in comparison with Stephenson and Fatoohi's calculations, however, and, because no more than three or four events of this kind take place every century, this alteration is not something to be given undue weight. The tilting of the axis, however, could produce new readings for the positions of celestial objects and affect all our astronomical data. The effect of such accidental events cannot be predicted by celestial mechanics.

A few computations were sufficient to assure me that this event had no effect on astronomical measurements. It produced a change about 1,000 times smaller than the power of our best telescopes to distinguish between two close points. I now knew that earthquakes, not even of the highest magnitude, can negate our astronomical records.

Since the end of the nineteenth century, astronomers have sought other possible solutions for the start of the Peloponnesian War in the interval 600 to 200 BC, without finding one to fit the descriptions provided by Thucydides. Historians suggested either that he was wrong or that he made the observation more to the north, in Thrace, where he owned several gold mine concessions. But, as he did not mention making any trips before the eighth year of the war, the last assumption is only conjectural.

Nobody extended the search for a different solution to a wider interval of time until the 1920s, when Nicolai

Morozov pointed out a sequence of eclipses that agreed with the observations: AD August 2, 1133; March 20, 1140; and August 28, 1151. In the 1970s Fomenko found another sequence: AD August 22, 1039; April 9, 1046; and September 15, 1057. In both cases the first eclipse was total in the Peloponnese. These solutions are the only ones that agree with Thucydides' descriptions.

Wanting to be certain of his source, Fomenko checked the translation of the original document with the linguist E.V. Alexeeva from the Faculty of Philology at the University of Moscow. She assured him that all the characteristics of the eclipse had been correctly interpreted.

In their 2001 paper, Stephenson and Fatoohi also touched on another aspect. In chapter 23 of Book One, Thucydides wrote that, during the war, "eclipses of the Sun occurred with a frequency unrecorded in previous history." Although observers can err when describing detail, they are less likely to make false statistical statements, in which the counting must be only approximately correct. Consequently, the two British astronomers redid the computations for all the eclipses that occurred twenty-seven years after and fifty years before 431 BC. Eight could be seen in the Peloponnese during the war (see figure 4.3) and sixteen in the preceding half century.

Stephenson and Fatoohi found Thucydides' remark baseless: eight events in twenty-seven years give a frequency similar to sixteen in fifty. But what can be said about Morozov's solution? From AD 1133 to 1160, seven eclipses had been visible from the Peloponnese, compared with fifteen in the previous fifty years. Again, the frequency is very much the same during both periods.

Things are different in Fomenko's case. Between AD 1039 and 1066, the Greeks could count eleven eclipses, while

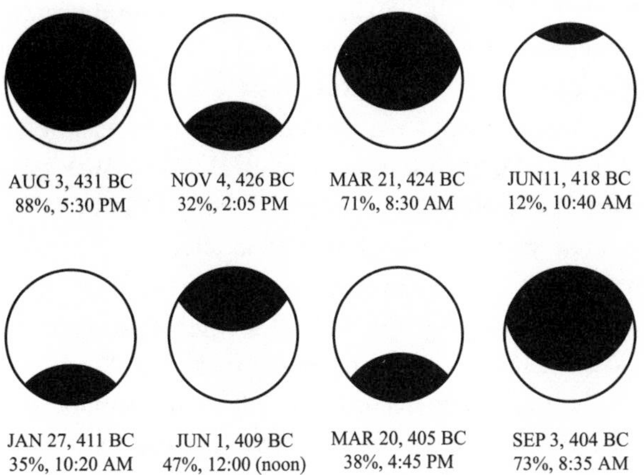

Figure 4.3—The eight eclipses that occurred in Athens between 431 and 404 BC, as computed by Stephenson and Fatoohi. Each of the eclipses is shown at its maximum phase. The percentage indicates the covered (dark) area.

during the preceding fifty years there were only fifteen. Eleven eclipses in twenty-seven years, compared with fifteen in half a century, give a significantly higher frequency, in accord with Thucydides. This fact seems to make AD 1039 more likely to mark the beginning of the Peloponnesian War.

I mentioned this fact in an email to Gleb Nosovski in late November 2003. A month later I received a new-year greeting from Anatoli Fomenko in which he congratulated me on my conclusion. But I didn't feel I deserved the compliment: my email to Nosovski had omitted my concerns.

Thucydides' high-frequency statement occurs in a paragraph claiming that the war took place during a period of disasters that had no previous match in history. Never had so many people been displaced, so many cities destroyed, and so much blood shed. Thucydides mentioned plague, drought, famine, and earthquakes of "unparalleled extent and violence." Could the remark about the eclipses be an exaggeration to emphasize his point?

It's also prudent to ask whether Thucydides' knowledge came from an expert source. A positive answer would favour the year AD 1039, while a negative one would not. Indeed, since small eclipses have a negligible effect on daylight, uninformed people are likely to overlook them.

Thucydides' words—"with a frequency unrecorded in previous history," or, in another translation, "at more frequent intervals than recorded at all former times"—made him seem familiar with astronomy, because he appears to rely on "recorded" sources. But in the original Greek the word Thucydides wrote is *mnemoneuomena*, which means "remembered" and doesn't imply methodical cataloguing.

Then again, why did Thucydides describe only two of the eight solar events of the war? Did he not see the others? That is likely: except for the 404 BC occurrence, the smaller eclipses had too little effect on the light of the day to attract attention. A description of the 404 BC eclipse is missing because the book failed to report on the last seven years of the war. Does this pattern mean that Thucydides relied on expert opinion?

Although eleven eclipses in twenty-seven years is above the average frequency in Athens, the length of the taken period is arbitrary. Extending the search by two years both in the past and in the future shows two new eclipses around the time of Scaliger's solution. This addition reduces the gap between traditional chronology and Fomenko because the increase of the time interval leads to no new eclipses in his case. Still, Fomenko's dating matches Thucydides' text better than the others'.

LIVY'S ECLIPSE

The eclipses of the Peloponnesian War are not Fomenko's only astronomical arguments against tradition. He also analyzed a solar eclipse described in Book 37 of Livy's *History of Rome*, a lunar eclipse mentioned by both Livy and Plutarch, and another lunar eclipse referred to in the New Testament at the time of Jesus' crucifixion. Fomenko disagreed with the accepted dates of March 14, 190 BC, for the first, June 21, 168 BC, for the second, and AD April 3, 33, for the third. His calculations placed all these events about a thousand years later.

Born Titus Livius in 59 BC, Livy was a Roman historian and a contemporary of Julius Caesar. His forty-five-volume work the *History of Rome from Its Foundation*, which survives from the original 142 books, is considered a jewel of world literature and fundamental for understanding the ancient era. Withdrawn from the political evils of his time, Livy led the quiet life of a man of letters. He died in AD 17, three years after Tiberius' accession to the throne.

In Book 37 he described an astronomical event: "When the consul [Publius Africanus] left for the war, during the games celebrated in honour of Apollo, on the fifth day before the ides of July, in a clear sky during the day, the light was dimmed since the Moon passed before the circle of the Sun."

This passage gives the date of the month when the eclipse took place. In the ancient Roman calendar, the days of March, May, July, and October 15, and those of the 13th for the other months, were called *ides*, so the "fifth day before the ides of July" is July 10.

In the spirit of tradition, Fomenko interpreted the text as saying that the contour of the Moon was visible, which happens only when the Moon passes below the centre of the Sun. This optical effect is known to astronomers, who noticed at the end of the nineteenth century that the eclipse of March 14, 190 BC, didn't satisfy this condition in the given geographical zone.

The traditional date was nevertheless unchallenged, probably because there is no viable solution close to 190 BC. Surveying all the eclipses from 600 BC to AD 1600, Fomenko found only one that matched the text: AD July 10, 967. But it is also possible that Livy's observation "the Moon passed before the circle of the Sun" has nothing to do with the visual effect mentioned above. It might only be the way the writer chose to portray the eclipse or to suggest that he was aware of its cause. Unlike the reference in the *History of the Peloponnesian War*, where Thucydides explicitly mentions the occurrence of stars, it is unclear if this description implies the passage of the Moon below the centre of the Sun. If it does, Fomenko's dating is correct. Otherwise, Livy's remark is unhelpful because it has too many solutions, including the traditional one.

THE LUNAR ECLIPSE OF LIVY AND PLUTARCH

Things were more straightforward for the lunar eclipse reported by both Livy and Plutarch. The latter lived from AD 45 to 125 and served as a priest in the temple of Apollo at Delphi, the site of the famous Delphic Oracle. Plutarch is mostly known for having written the biographies of many prominent men of antiquity, including Pericles, Alexander the Great, Pompey, and Julius Caesar. In his *Life of Aemilius Paulus*, Plutarch described the following event: "When it was night and, supper being over, all were turning to sleep and rest, all of a sudden the Moon, which was then fully high in the heavens, grew dark and, gradually losing her light, passed through various colours, and at length was totally eclipsed."

From this and other contextual information given by Livy and Plutarch, researchers concluded that a lengthy total lunar eclipse occurred on the night of September 4 to 5 (Roman calendar) of an unknown year, after the summer solstice. Traditional chronology computed the Julian date of June 21, 168 BC, but this day fell before the summer solstice, in violation of the historical text.

Attempts to find a viable solution between 300 and 100 BC proved fruitless. Fomenko extended the search to the interval from 600 BC to AD 1600 and found three candidates, one in each of the years AD 415, 955, and 1020. By also including the eclipses that took place at sunset, he discovered another one in AD 434. Those from 955 and 1020 had the longest duration, in agreement with Livy and Plutarch, and both were more likely than the others. Fomenko provided no historical reason to prefer either one, but pointed out that both confirmed his millennium shift.

Still, he wanted more evidence for his theory, so he researched other events.

THE CRUCIFIXION

The dating of the eclipse chronicled in the New Testament at Jesus' crucifixion turned out to be more problematic than the others. Three passages refer to it. Matthew 27:45 reads: "Now from the sixth hour there was darkness over all the land unto the ninth hour." Mark 15:33 merely rewords the statement in Matthew: "And when the sixth hour was come, there was darkness over the whole land until the ninth hour." Luke 23:44–45 provides additional information: "And it was about the sixth hour, and there was darkness over all the earth until the ninth hour. And the Sun was darkened, and the veil of the temple was rent in the midst."

According to tradition, this eclipse was lunar and occurred on AD April 3, 33. That is surprising in view of Luke's remark about the Sun, which suggests a solar eclipse. But John 19:14 states that the crucifixion took place around the time of Passover: "And it was the preparation of the Passover, and about the sixth hour: and he saith unto the Jews, Behold your King!"

The computation of Passover is based on a lunar calendar and requires a full moon, which can take place only if the Earth is between the Moon and the Sun. For a solar eclipse to happen, the Moon must be between the Earth and the Sun. So the passage in Luke probably means that the eclipse occurred at night, unless the Passover information is disregarded.

The main problem is that the AD April 3, 33, event had a short duration in and around Jerusalem, lasting only a few minutes instead of about three hours—as the Bible implies. In the late 1920s Nicolai Morozov pointed out the existence of a long-lasting lunar eclipse that met all the biblical criteria. The only such eclipse between 200 BC and AD 800 took place on AD March 21, 368.

As in other cases, Fomenko extended the search up to AD 1600 and discovered another solution: AD April 3, 1075. Though this eclipse was as short as the traditional one, Fomenko preferred it because it had all the characteristics of the latter. But both choices are worse than Morozov's, which matches the ancient description. So what source is to be accepted: the text, the tradition, or the millennium-shift hypothesis? And, although Morozov's solution accords with the biblical account, how does AD March 21, 368, agree with the dating of other astronomical events at the time of Jesus' birth?

CALENDAR REFORM AND THE COUNCIL OF NICAEA

The date of Jesus' birth is also controversial. Its first computation is attributed to a sixth-century monk, Dionysius Exiguus (the Little)—from what is now Dobrogea, on Romania's Black Sea coast—who became known for his mathematical skills and vast knowledge of astronomy. In 525, at the request of Pope John I, he wrote *Liber de Paschate*, a table of Easter dates with instructions on how to calculate them. As a starting point he took the year AD 1, about which he indicated:

> If you want to find out which year it is since the incarnation of our Lord Jesus Christ, compute fifteen times 34, yielding 510; to these always add the correction 12, yielding 522; also add the indiction of the year you want, say, in the consulship of Probus Junior, the third, yielding 525 years altogether. These are the years since the incarnation of the Lord.

We can only speculate what Dionysius meant in this cryptic paragraph. But nobody disputes that the division of time beginning with AD 1 came into existence with this manuscript. Although the legitimacy of the dating has been questioned ever since, nobody before Fomenko placed Jesus a millennium later than AD 1. An apparently simple way to contradict Fomenko is to invoke the sixteenth-century calendar reform of Pope Gregory XIII, whose ten-day correction allows the dating of Caesar's reign. But, as I will further explain, Fomenko and Nosovski found this approach far from convincing.

Our calendar has its origins in 46 BC, when Julius Caesar established the length of the months at thirty or thirty-one days and introduced the leap year. The emperors Augustus, in 8 BC, and Constantine the Great, at the time of the Council of Nicaea, made small changes to this calendar.

The Julian year differs from the solar year by a few minutes and, as a result, timekeeping errors added up. By the sixteenth century the beginning of spring fell in early March. So, in 1582, Pope Gregory XIII acted on the advice of the German mathematician and astronomer Christopher Clavius and shortened that year's month of October by ten days. He also changed the rules of the leap year, cancelling February 29 in the years ending in two zeroes, except for the multiples of 400, such as 1600, 2000, 2400, and so on.

There are documents that connect Jesus to Tiberius, the Roman emperor from AD 14 to 37. In his *Annals*, Cornelius Tacitus (AD 55–120) wrote that Christians "derived their name and origin from Christ, who, during the reign of Tiberius, had suffered death by the sentence of the procurator Pontius Pilate."

Tiberius was the second Roman emperor, born soon after Julius Caesar's first calendar reform, and he can be easily related to the later reformers Augustus and Constantine. These connections mean that, given the ten-day correction of Pope Gregory XIII, one can calculate with some accuracy when Tiberius lived. Even if it were impossible to pin down the exact years, it would allow one to show that the error cannot be counted in hundreds of years, as Morozov and Fomenko claimed.

But Gleb Nosovski disagreed: the corrections made to

the Julian calendar were wrong. He objected to the computations of Luigi Lilio (Aloysius Lilius), a Naples physician, astronomer, and mathematician, on whose results Clavius had recommended the changes. Nosovski started from the papal bull given on February 24, 1582, which stated:

> Our care was not only to reinstate the equinox in its long ago nominated place from which it has deviated since the Council of Nicaea by approximately ten days, and to return the 14th Moon [full moon] to its place, from which it has deviated by four and five days, but also to settle such modes and rules according to which future equinoxes and the 14th Moon would never move off their places . . . Therefore, to return the equinox to its proper place established by the Church fathers of the Council of Nicaea on the 12th day before the April calends [March 21], we prescribe and order relative to October of the current year, 1582, that ten days, from the third day before nonas [October 5] to the eve of the ides [October 14] inclusive, be deleted.

Nosovski noticed two errors in the bull. The first concerns the time difference between spring equinox and full moon, which the text claimed would be kept constant in the future. This is impossible because the cycle of full moons and the date of the equinox shift at different rates. But probably this mistake rests with those who drafted the bull, for it is hard to believe that an astronomer made it.

The second error, however, is important, and has to do with determining the equinox and the full moon. To understand Nosovski's objection and the way he solved the problem, recall the First Ecumenical Council of Nicaea, which Scaliger dated at AD 325 (see chapter 2). This year is essential for the accuracy of the Gregorian reform; the ten-day correction depends on it.

The Christian calendar consists of a rigid part and a flexible part. The former is the "old-style" solar Julian system, with its fixed celebrations, whereas the latter is the lunar *Easter Book*, which determines the variable Christian feasts and festivals. All religious services are based on these two systems.

The difficulty of combining the lunar and the solar calendars has faced theologians ever since the Christian Church began to celebrate Easter—which, according to tradition, started in the second century AD. The rules were given in the *Easter Book*, canonized by the Council of Nicaea in AD 325. But this dating remains unclear because the original text of the Nicaean Creed has not survived. The only existing document that tells how to compute the celebrations is the message of Constantine to the bishops who were absent from the council, and it doesn't mention that the Orthodox Easter should take place after the equinox.

In the fourteenth-century *Collection of Rules of the Holy Fathers of the Church* by the medieval scholar Matthew Vlastar, the conditions for determining the anniversary of Christ's resurrection are described as follows:

> The rule on Easter has two restrictions: not to celebrate together with the Israelites and to celebrate after the spring equinox. Two more were added by necessity: to have the festival after the very first full Moon after the equinox and not on any day but on the first Sunday after the full Moon. All the restrictions except the last one have been kept firmly until now, but now we often change for a later Sunday. We always count two days after the Passover [full moon] and then turn to the following

> Sunday. This happened not by ignorance or inability of the Church fathers who confirmed the rules, but because of the lunar motion.

So, by approximately AD 1330, when Vlastar wrote his account, the last condition of Easter was violated: if the first Sunday happened to be within two days after the full moon, the celebration of Easter was postponed until the next weekend. This change was necessary because of the difference between the real full moon and the one computed in the *Easter Book.* The error, of which Vlastar was aware, is twenty-four hours in 304 years.

The *Easter Book*, therefore, must have been written by AD 722 (722 = 1330 – 2 x 304). Had Vlastar known of the *Easter Book*'s AD 325 canonization, he would have noticed the three-day gap that had accumulated between the computed and the real full moon in more than a thousand years. So he either was unaware of it or knew the correct date, which could not be near 325.

With the Easter formula derived by Karl Friedrich Gauss in the nineteenth century, Nosovski calculated the Julian dates of all spring full moons from the first century AD up to his own time and compared them with the Easter dates obtained from the *Easter Book.* He reached a surprising conclusion: three of the four conditions imposed by the Council of Nicaea were violated until 784. When proposing the year 325, Scaliger had no way of detecting this fault because, in the sixteenth century, the full-moon calculations for the distant past couldn't be performed with precision.

Another reason to doubt the validity of 325 is that the Orthodox Easter dates repeat themselves every 532 years. The last cycle started in 1941. The previous ones were 1409 to 1940, 877 to 1408, and 345 to 876. But a periodic process is similar to drawing a circle—you can choose any starting point. Therefore, it seems peculiar for the council to meet in 325 and begin the Easter cycle in 345.

Nosovski thought it more reasonable that the First Council of Nicaea had taken place in AD 876 or 877, the latter being the starting year of the first Easter cycle after 784, when the *Easter Book* was probably compiled. This conclusion agreed with his calculations, which showed that the real and the computed full moons occurred within one day only between AD 700 and 1000. From 1000 on, the real full moons occurred more than twenty-four hours after the computed ones, whereas before 700 the order was reversed. The years 784 and 877 also match the traditional opinion that about one century passed between the compilation and the canonization of the *Easter Book.*

Nosovski didn't stop here. He also wanted to confirm the date of Jesus' resurrection. For this calculation he used the conditions that had guided Dionysius the Little in his determination of the First Easter. Nosovski obtained them from the same source as Scaliger: the fourteenth-century writings of Matthew Vlastar. With the help of a computer, the Gauss formula, and the *Easter Book* rules, he checked the interval from 100 BC to AD 1700. Only the year 1095 satisfied all conditions, though nine others fulfilled the basic requirements that the full moon was on a Saturday, March 24, and the resurrection took place on a Sunday. But none of them was within the acceptable margins of traditional chronology.

One argument against the year 1095 is its derivation from medieval sources. But these are the same sources employed by Scaliger, who, Nosovski explained, drew the wrong conclusion because of the imprecise astronomical methods of the sixteenth century. Another argument against

Nosovski is that Jesus lived some three centuries after the Council of Nicaea, which decided to put him on an equal footing with God the Father. Nosovski recognized the difficulty of providing a good reason for this reversal. At the same time, he claimed to have contradicted not the truth but the point of view of the Church's history, which was formed only about five centuries ago.

Still not satisfied with having reversed the two events, Nosovski re-examined his dating of the Nicaean Council in a more recent article and suggested the year AD 1343. His new "proof" is linguistic, claiming that Nike was the Greek goddess of victory and, therefore, the Council of the Victors is the same as the Council of Nicaea. For reasons that will be made clear in chapter 8, this argument is not convincing. The years 876 or 877 remain his best solutions.

Another surprising statement in Nosovski's work is his identification of Dionysius the Little with Matthew Vlastar. But this claim is based on an attempt to explain how Dionysius arrived at the computation rules given in *Liber de Paschate*, and, as I mentioned earlier, one can only guess the source of those algorithms.

Both Fomenko and Nosovski have remarked that Scaliger opted for a more distant date whenever he had a choice. For this, they blamed the spirit of the time, which claimed "the older, the better." As an example, Nosovski showed why the year AD 325 was the earliest Scaliger could assign for the First Nicaean Council. According to ancient documents, in 325 the spring equinox couldn't occur later than March 21, and the first time it fell on that date was at the end of the third century. That is one of Fomenko and Nosovski's many arguments against Scaliger's chronology.

JESUS

In his article on the Council of Nicaea, Nosovski took a different approach to the dating of Jesus' life. He first interpreted the star of Bethlehem as a famous supernova, which gave birth to the Crab Nebula recorded by Chinese observers in AD 1054. Then he argued that the biblical passage in Luke quoted earlier describes the total solar eclipse observed in the Mediterranean area on February 16, 1086. So Jesus might have been born in 1054 (or perhaps 1053) and crucified in 1086.

But the dating of the 1054 supernova has been disputed in literature. In 1997 the Italian Giovanni Lupoato noted that the *Rampona Chronicle* may have marked the explosion. To Lupoato, the recorded date of June 24, 1058, was a transcript error in the fifteenth-century copy that survived the original: MLVIII (1058) instead of MLIIII (1054). Still, it is not clear if this chronicle reports the same event. Moreover, since Fomenko and Nosovski also challenged the Chinese chronology, the year 1054 might be incorrect even in terms of their own standards.

Fomenko and Nosovski gave three possible dates for the resurrection of Jesus: 1075, 1086, and 1095. The fact that these dates were all close to each other, Nosovski believed, suggests that Christ lived in the eleventh century AD. Each year was obtained by looking at different documents, all of which are veiled in legend. Therefore, some rare astronomical events that happened near the time of Jesus' life might have been associated with his birth or crucifixion.

How credible are these conclusions? The 1086 date is the least likely one, for reasons made clear above. The year 1075 is based on a lunar eclipse satisfying most characteristics described in the Bible, except for the long duration; though better than the AD 33 choice, it is worse than Morozov's AD 368 solution. The year 1095 has a weak point too: the Council of Nicaea precedes it. So this date isn't more likely than the previous ones, unless church history has got it all wrong.

The dating of Jesus' life, however, is a questionable landmark because all the existing sources referring to it are tainted with legend. Some authors, such as theologians Alvin Boyd Kuhn and Tom Harpur, go so far as to claim that Jesus never existed and that the early church borrowed his story from ancient religions. Such a wide spectrum of contradictory beliefs and evidence makes it hard to rely on the life of Jesus for chronological purposes.

The situation changes, however, with the Council of Nicaea. Nosovski appealed to the same sources Scaliger had and arrived at the year 876 or 877 by using verifiable arguments. This date seems worthy of further explorations, which might endorse or disprove its validity.

But these conclusions are not the only ones Fomenko and his collaborators base on astronomy. Their methodology also mixes astronomical records with celestial mechanics and mathematical statistics. Applying these techniques has led them down some intriguing paths.

CHAPTER 5

The Moon and the *Almagest*

When I follow the windings of heavenly bodies, I no longer touch the earth with my feet, but stand in the presence of Zeus and take my fill of ambrosia—food of the gods.

CLAUDIUS PTOLEMY

No science is 100 percent exact: each of them involves certain approximations. But if there is one science that comes closest to perfection, it's celestial mechanics. From 1687, when Newton inaugurated this field of study, until today—the age of electronic computers—celestial mechanics has been able to forecast the return of comets, to discover new planets, and to guide rockets in space. No mission has ever failed because of it, no eclipses have ever occurred at any times other than those it predicted.

As I showed in chapter 1, Velikovsky can be refuted within the framework of celestial mechanics by calculations that comprehend everything seen in nature. If we don't trust these results, and prefer instead to accept the existence of unobserved forces, we might as well believe in fairy tales.

Astronomical data and historical texts that contradict celestial mechanics are, therefore, highly suspect to any reasonable mind. That was Fomenko's point when he began his research in chronology, and few people would disagree with this approach. The conclusions, however, are startling.

CELESTIAL MECHANICS

The question that started Fomenko on the path of historical dating had nothing to do with history. When he attended Mikhail Postnikov's lectures on Morozov in the early 1970s, he was studying the motion of the Moon.

The Moon's orbit around the Earth has concerned many researchers since the time of Isaac Newton. But not until the end of the nineteenth century, when the American mathematician George William Hill found a suitable model, did experts have a reasonable understanding of it. Still, some details of the orbit remain unclear, and Fomenko was dealing with one of them.

Fomenko looked at what researchers call the Moon's elongation, which is the angle between the Moon and the Sun as viewed from the Earth. The change of the speed at which this angle varies defines the acceleration of the Moon's elongation, denoted by the expression D" (read D double prime). This quantity is computable from observations, and its past behaviour can be determined from records of eclipses.

The values of D" are very small (measured in seconds of arc per century squared), and most of them range between –18 and +2. The acceleration is slightly above zero and almost constant from about 700 BC to AD 500, then drops significantly for the next five centuries, and finally settles around –18 after AD 1000 (see figure 5.1). The problem,

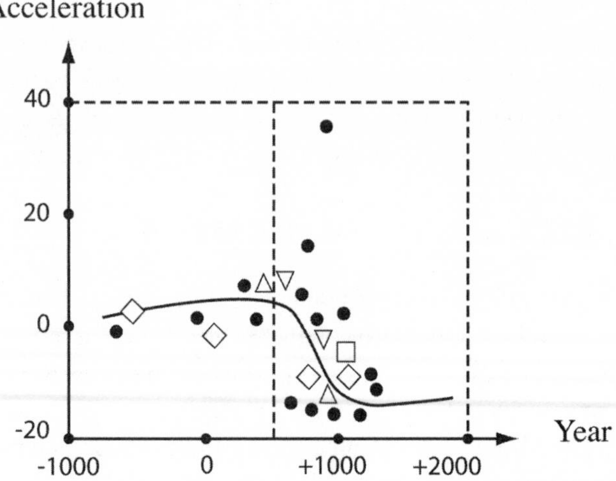

Figure 5.1—The graph of D" obtained by Robert Newton. The symbol • indicates values calculated from recorded solar eclipses, ◊ lunar eclipses, Δ solar eclipse duration, and ∇ phases of solar eclipses.

however, is that this variation cannot be explained on the basis of gravitation, according to which the graph should be a horizontal line.

In 1979 Robert R. Newton, a professor at Johns Hopkins University in Baltimore, published the first volume of *The Moon's Acceleration and Its Physical Origin*, which considered the issue by looking at historical solar eclipses. The second volume appeared in 1984 and dealt with the same problem from the point of view of lunar observations. Unable to explain the behaviour of D", Newton concluded that some unknown forces must be responsible for its variations.

In January 1980, when Fomenko submitted his conclusions on the Moon's acceleration to a respected journal in

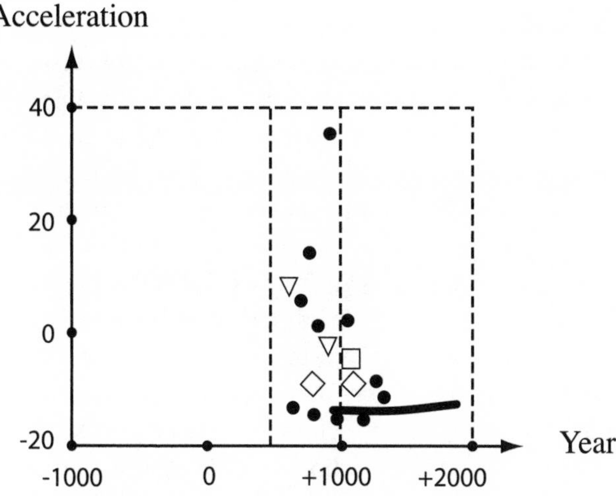

Figure 5.2—Fomenko's graph of D" after he shuffled the calendrical dates of the astronomical events considered by Robert Newton. Until about AD 500 there are no observations, and those from 500 to 900 are unreliable, whereas the ones between 900 and 1900 lead to an almost constant lunar acceleration.

celestial mechanics, he had read Robert Newton's first volume and several of his articles. The Russian mathematician rejected the idea of unknown forces. Using his new chronology, he redated Newton's astronomical records and obtained the graph in figure 5.2, where D" was almost constant. This result agreed with gravitational theory, according to which a deceleration of the Moon corresponds to a slowdown of the Earth's rotation around its axis.

Newton was probably ignorant of Fomenko's results; his second volume, published in 1984, didn't mention them. Instead, he continued to present evidence for the unpredictable changes of the Moon's acceleration until he died in 1991 at the age of seventy-two, without ever hinting he had read Fomenko's work.

Among the possible non-gravitational factors that change the values of D", Newton proposed the tidal friction between water and sea bottoms, the Earth's magnetic force, the withdrawal of the ice caps, and the growth of the Earth's core. But he couldn't tell how realistic these proposals were, since there is no way at present to compute their contribution to the Moon's acceleration. It is far from clear that their influence on D" could account for the bend portrayed in figure 5.1.

The two researchers approached the problem from radically different starting points. Newton didn't question the traditional dating of any eclipse, whereas Fomenko doubted them all. Newton was choosy when it came to ancient descriptions, trusting only a handful of the 370 cases he studied. In the second volume of his book he complained: "We have found too many instances of an eclipse that could not possibly have been total but that was so recorded, sometimes in a quite picturesque manner." Fomenko, instead, took the word of the ancients for granted, as he did with Thucydides. Depending on the choices they made, the two experts reached different conclusions.

Fomenko appears to be more convincing because he resorted to no mysterious forces. But a closer look at figures 5.1 and 5.2 shows that his graph after AD 900 is very much the same as Newton's. In the middle period, for which Newton noticed a sharp drop of D", Fomenko obtained a mixture of results, which he deemed unreliable. The most ancient period vanished because his chronology is shifted forward.

Ignoring the period before AD 900 means there is no significant difference between the results of Newton and Fomenko. The change of calendrical dates has not led to an approximately straight line starting with antiquity but only eliminated the data before AD 500 and created confusion between the years 500 and 900. This result doesn't prove that Fomenko's new chronology is correct. In fact, he characterized the fluctuating values of D" as follows:

> Either the scarce astronomical descriptions that chronologists ascribe to this period [before AD 900] are very nebulous, or, what is more probable, these chronicles are also misdated, and the events they describe are in need of redating. However, due to utter vagueness of the remaining astronomical descriptions, they cannot be used for dating purposes since they offer too many solutions.

Though this result doesn't support Fomenko's new system, it challenges tradition. Newton's contribution can be viewed in much the same way: in the absence of unknown forces, his graph casts a doubt on conventional chronology. That leaves three possibilities: there are unknown forces; the laws of physics have changed since the past millennium; or traditional chronology is wrong.

The first two choices are not impossible, but physicists would argue that they seem very unlikely. The controversies described in previous chapters support the idea that the timetables of history are fragile, so the balance of probabilities tilts in the direction of accepting the third possibility. Fomenko's result shows the need for taking a closer look at traditional chronology.

ALLEGED FABRICATIONS

The Moon's acceleration was not the only issue on which Newton and Fomenko disagreed. They also differed on the most influential astronomy book ever written: the *Almagest.*

Its author, Claudius Ptolemy—considered one of the greatest scientists of antiquity, particularly in his contributions to geography, optics, and mathematics—lived during the second century AD mostly in Alexandria, on the southeast shore of the Mediterranean Sea.

Ptolemy wrote the *Almagest* during the reign of the Roman emperor Antoninus Pius, which is traditionally set from 138 to 161. Any firm evidence for a different dating of this treatise would change the entire chronology of Rome. Divided into thirteen books, the work touches on the main problems of astronomy, from the nature of the universe to lunar and planetary motion. It also contains detailed star catalogues and records of eclipses, occultations, and equinoxes. The original version of the *Almagest* has been lost, but in its many translations the work has been in circulation for almost two millennia.

In 1977 Robert Newton published *The Crime of Claudius Ptolemy*, a book in which he accused the ancient astronomer of the greatest sin a scientist can commit—fabricating evidence. Newton argued that many coordinates presented in the *Almagest* as observations were nothing but fraud.

He started with Ptolemy's records of equinoxes and solstices for AD 132, 139, and 140. The ancient scientist had given times of occurrence that were wrong by twenty-eight to thirty-six hours. This is a *huge* discrepancy because, on the one hand, equinoxes and solstices advance by only twenty minutes a year (see chapter 2); on the other, Aristarchus, Euctemon, and Hipparchus, who lived centuries before Ptolemy, made less than seven-hour errors. Newton proposed schemes according to which Ptolemy's results had been produced from earlier observations.

Newton's list of alleged fabrications includes coordinates of eclipses, planets, and stars. To Newton, the goal of

Ptolemy's deliberate fraud was to make data agree with theory, as immature students might do to cover the tracks of their weak laboratory work. When suggesting this metaphor, Newton didn't seem to worry about the difference between a student paper, which ends up in the recycling bin, and a book that is open to the scrutiny of peers.

But criticism of Ptolemy isn't new. As early as AD 1008 the Egyptian astronomer Ibn Yunis remarked that the positions recorded in the *Almagest* contained serious errors. In 1817 the French scientist Jean Baptiste Joseph Delambre preceded Newton by asking: "Did Ptolemy do any observing? Aren't the observations he claims to have made mere computations from his tables and examples to help him explain his theories?" Closer to the present, the dissertation of J.P. Britton, defended at Yale in 1967, claimed that the equinoxes and solstices of the *Almagest* had been forged.

In 1979 Rolf Brahde of the Institute of Theoretical Astrophysics in Oslo wrote a rave review of Newton's *Crime of Claudius Ptolemy*, one the historian of science K.P. Mosegaard endorsed. A few years later the Dutch mathematician and Zurich professor Bartel Leendert van der Waerden went as far as to state: "Delambre and Newton have convincingly proved that Ptolemy had systematically and intentionally falsified his observations in order to make his results agree with his theory."

But not everyone shared this view. In 1978 the journal *Science* published the appraisal of University of Pittsburgh professor Bernard Goldstein, who suggested that Robert Newton had often distorted the facts. A year later Owen Gingerich of Harvard University, Noel Swerdlow of the University of Chicago, and Victor Thoren of Indiana University, after whom an asteroid is named, remarked in *Scientific American* that Newton's conclusions were based on flawed statistical analysis that disregarded the methods of early astronomy. In 1988 the respected translator of the *Almagest* and Brown University historian of science Gerald J. Toomer took a similar position, dismissing Newton's arguments.

Among the more recent critics were Gerd Graßhoff, from the University of Bern, and Oscar Sheynin of Berlin. While Graßhoff found the arguments against the ancient astronomer superficial and unjustified, Sheynin wrote that Ptolemy didn't falsify but merely selected convenient observations, a practice common in ancient times. In the opinion of James Evans of the University of Puget Sound in Tacoma, Washington, very few historians of astronomy have agreed with Newton's fabrication theory.

But why regard Newton's work as either black or white? Could it have been correct in some respects and wrong in others? These were the questions that Fomenko thought were worth pursuing.

A DIFFERENT OPINION

Fomenko considered *The Crime of Claudius Ptolemy* a valuable piece of work but disagreed with its conclusions. Being familiar with Nicolai Morozov's seven-volume *Christ* (see chapter 2), the Russian mathematician knew things the Western critics did not. In the 1920s Morozov had already thought about the *Almagest* and had offered his own opinion: Ptolemy, or whoever produced his catalogue, must have lived much later than the second century AD. In Fomenko's view, Newton confirmed this claim without knowing it.

From ancient times astronomers have used two systems of coordinates to record observations: equatorial and ecliptic. As the name implies, the first employs the equatorial plane as reference, whereas the second relies on the ecliptic (the plane of the Earth's orbit around the Sun). Before the seventeenth century, both systems were in common use; after that, the equatorial coordinates became dominant for being more convenient and for leading to more precise observations.

Ptolemy preferred ecliptic coordinates. In the Middle Ages, astronomers deemed them more reliable because the equatorial plane oscillates owing to precession (see figure 2.2). What they did not know was that, because of the planets' gravitational pull, the ecliptic varies too—though much less than the equator—so these coordinates offer no real advantage. Ptolemy was also unaware of that.

For his astronomical observations, Ptolemy used an astrolabe, which he duly described in the *Almagest.* This instrument is not easy to handle, and it might have been a source of his errors. Some researchers think he made some equatorial measurements and transformed them into ecliptic. The corresponding formulae depend on the precession rate—roughly 1.4 arc degrees per century; Ptolemy used 1 degree instead. This calculation led him to new errors, but they are consistent and therefore easier to trace.

From the start, Morozov was surprised to see Ptolemy's star catalogue claiming a 10 arc-minute precision. Such good accuracy could not be achieved without a minute hand clock, which was not invented until much later, in the Middle Ages. So, even before proceeding with his mathematical analysis, Morozov suspected a medieval origin for the *Almagest.*

One way to determine the year in which an observation was made is to use precession. Since the Earth's axis rotates 360 degrees in 26,000 years, the angle, and therefore the longitude of every star, changes by about 50 arc seconds per year. Dividing the difference between the present equatorial longitudes and those of Ptolemy by fifty, Morozov obtained an apparently shocking result: Ptolemy's observations came from the sixteenth century AD.

But Morozov didn't jump to conclusions. This anomaly could be due to the error Ptolemy had introduced by taking a precession rate of 1 degree instead of 1.4. Another possible reason was the star catalogue, which came from a sixteenth-century Latin translation. So Morozov checked the Greek version and learned that the stars' longitudes differed by about 20 degrees, and therefore corresponded to the traditional second-century dating.

He thought that perhaps the coordinates of the Latin translation had been updated. But a thorough comparison of the two editions showed the observational errors in the Greek edition to be much smaller than those in the Latin text. This information contradicted the common assumption that the Greek version was an early copy and not a later translation.

Morozov noticed other strange things about the Greek edition. First, the records corrected the observations of the stars' latitudes with the value of the atmospheric refraction, a phenomenon discovered only in the Middle Ages. Second, due to precession, the choice of Polaris (α-Ursae Minoris) as the first star of the Greek catalogue would not have made sense in the second century AD, when the axis of the Earth pointed closer to β-Ursae Minoris. Last, but certainly not least, though present in the catalogue, the star Achenar had been visible in Alexandria only since the fifteenth century AD.

These anomalies made Morozov conclude that the *Almagest* had been written at least a millennium later than historians thought. But was this claim legitimate?

IN SEARCH OF A SOLUTION

Fomenko wasn't convinced by Morozov's arguments. He found them subjective and too dependent on the editions Morozov consulted, texts that could have differed from the original. Nevertheless, he also thought something was wrong with the dating of the star catalogue and decided to study it himself. The best version turned out to be the one by Christian Peters and Edward Knobel, published in 1915. The authors had compiled the star positions from all historical manuscripts, adding many useful comments and remarks. As Gerald Toomer confirmed in his translation of the *Almagest*, the work of Peters and Knobel is the only comprehensive source.

Fomenko noticed a serious problem with the claim that the 20-arc-degree gap between the stars' longitudes in the Greek and the Latin versions guaranteed Ptolemy's existence in the second century AD. The founders of the modern dating system had tacitly assumed for both the ancient and the medieval longitudes the same "starting location," the astronomical equivalent of the Greenwich meridian. Fomenko found examples in which this presumption led to gross mistakes.

One of them concerned the *Theatrum Cometicum* (Theatre of Comets), the work of the celestial cartographer Stanislaw Liubeniecki, published in Amsterdam in 1668. Assuming its star longitudes had the same starting location as the star longitudes of the *Almagest*, Fomenko was led to a sky configuration from the fifth century BC—an error of more than two millennia. This mistake explained why, without knowing the reference point, the precession phenomenon couldn't explain the origin of Ptolemy's catalogue and provided an argument against its traditional dating.

Fomenko had an idea of how to solve the problem. His attempt was based on the fact that every star has a proper motion that is unrelated to the apparent one due to precession. The discovery of this phenomenon is attributed to Edmund Halley at the beginning of the eighteenth century. Ptolemy had also asked if stars moved independently of each other, but he erroneously answered the question in the negative.

Ptolemy's error was certainly understandable. The motion of stars can be detected only by hundreds of years of precise observations. It is similar to that of an airplane being watched from afar, whose flight can be deconstructed into a radial component (away from or towards the observer) and a tangential one, which can be actually observed. While the radial component has no effect on the values of celestial coordinates, the tangential one modifies both the longitude and the latitude of a star.

The tangential component is responsible for the shapes of constellations, whose changes can be determined today. Figure 5.3, for example, shows the Big Dipper in the years AD 2000 and 100,000. Using the relative positions given in the *Almagest* and comparing them with the present ones, Fomenko decided to find out when the book had been written. But that goal was not so easy to achieve.

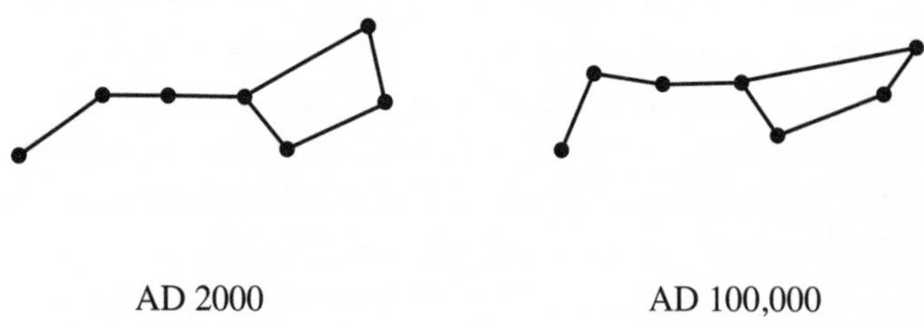

AD 2000

AD 100,000

Figure 5.3—The Big Dipper (Ursa Major) as seen from the Earth in AD 2000 and AD 100,000.

One impediment was the use of Ptolemy's catalogue for tracing the motion of some stars. If the catalogue's dating was incorrect, the computed speeds of these stars were also wrong. Fomenko had, therefore, to trace the history of those determinations and eliminate from his analysis the stars related to the *Almagest.*

But the most difficult process was to identify the catalogued stars. Since the sixteenth century this problem had preoccupied many famous astronomers, and the edition Fomenko consulted seemed to have found the answers. Still, he wasn't sure that was the final word in the matter, and he had good reasons for doubt.

In ancient and medieval times the shapes of constellations were not standardized and their description was often vague. Therefore, telling which star from the catalogue corresponds to the one you see in the night sky is difficult. Ptolemy provided positions and magnitudes. For the brightest objects the identification is easier, because there are few to choose from, but with the fainter stars things get complicated: their coordinates and also their magnitudes in the *Almagest* are often wrong.

Research done on this problem assumed observations made in the second century AD, a fact that influenced the identification. For different suppositions the outcome changes, because the date of the record is crucial for computing the speed of stars. A researcher who wants to date Ptolemy runs the risk of having to assume that the catalogue was written, say, in AD 600 in order to prove that the ancient astronomer lived around AD 600. This circularity is of no use to either a mathematician or a historian.

Fortunately, identification is not so difficult for most stars. Those of the zodiacal constellations, for example, pose fewer problems because they have been studied more carefully for astrological purposes and there is more historical information about them. Of the 350 zodiacal stars recorded in the *Almagest*, Fomenko chose to focus only on the very fast ones—those with an individual motion of at least 1 arc second per year. He made this decision because slower objects could have travelled distances less than those resulting from Ptolemy's errors.

Eight stars satisfied the high-speed condition. In fact there was a ninth, but its 5-degree error corresponds to its own motion in 5,000 years, so Fomenko dropped it from the list. Among the eight stars, only the brightest—Sirius, Arcturus, and Procyon—were certain. For the others, different authors had proposed different identifications.

For a rough dating of Ptolemy's record for each star, Fomenko applied the method of "least squares." In geometric terms, this method consists of taking a perpendicular from the recorded position in the catalogue to the line on which the

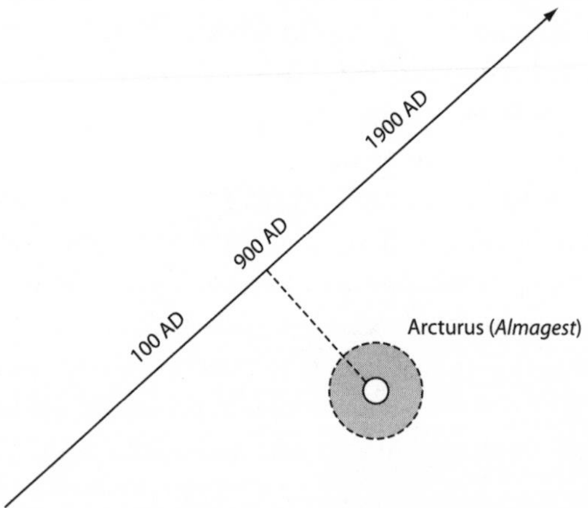

Figure 5.4—The application of the method of least squares to the dating of the star Arcturus.

star has moved in reality. The foot of the perpendicular is the closest point between the line and the star, so the dating of this position is the most likely one (see figure 5.4).

The method failed. It gave the years AD 500 for Sirius, 800 for Arcturus, and 1850 for Procyon according to one identification and 1450 for Procyon in another. The remaining five stars yielded widely diverging dates too, from 2000 BC to AD 1800. This result was not encouraging, and Fomenko thought of another way to attack the problem.

With his associates Gleb Nosovski and Vladimir Kalashnikov, he considered all the stars simultaneously. Their idea was simple. Take the distance between the position of a star as recorded in the *Almagest* and its real position in a given year as determined by computations. For that year and for all stars, sum up those distances. Repeat the procedure for all years in some interval long enough to avoid bias, say 500 BC to AD 1800. Compare the results and choose the year for which this sum is the smallest.

This method provides an optimal estimate. It doesn't take the year at the foot of the perpendicular, as in figure 5.4; instead, it connects the *Almagest*'s position of the star with another particular point on the line. The new segment is usually longer than the perpendicular, but, taking all the stars and summing up the distances corresponding to the same year, the final result is minimal when compared with any other choice.

Though this minimum principle is standard in mathematics, its application to chronology could be deceiving. A straightforward attempt led to different results depending on the group of stars considered. The variations were as large as the ones obtained with the method of least squares. Therefore Fomenko and his colleagues thought of improving their results with a thorough error analysis. Estimates

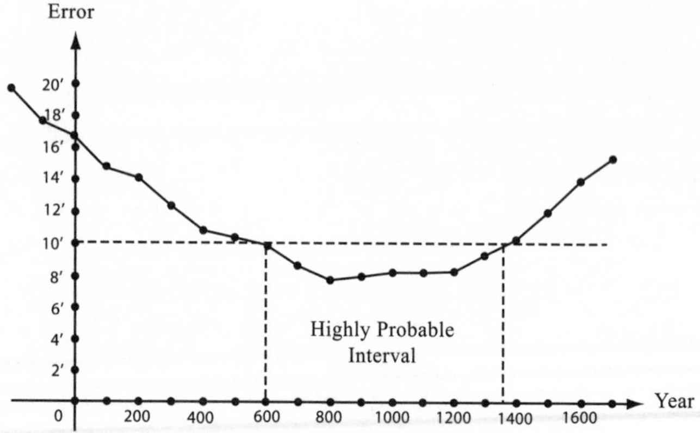

Figure 5.5—Fomenko's statistical analysis showed that the only time interval in which Ptolemy's errors are smaller than 10 minutes of arc is between AD 600 and 1300. The graph depicts the size of the errors from 200 BC to AD 1700.

for each century (see figure 5.5) pointed out that the only interval in which the errors were smaller than Ptolemy's 10-arc-minute precision was from AD 600 to 1300.

Of course, this conclusion depends on several assumptions, and the Russian mathematicians checked the reliability of the result. The estimate showed a very small but non-zero probability that the *Almagest* had been written outside this interval. It is like saying that, within the stated margin of error, an opinion poll is accurate nineteen times out of twenty. In other words, they had obtained a likely result, though not an absolutely certain one.

With admissible but not realistic changes in the parameters, the interval could have been extended back in time as far as AD 350, still two centuries away from the traditional dating. The good news was that the outcome didn't change if the mathematicians slightly varied the data. To gain more confidence in this procedure, they also tested star catalogues of the sixteenth and seventeenth centuries, as well as some computer-generated ones. The results proved more than satisfactory: they recovered the known dates within a ten-year margin of error.

OCCULTATIONS

The Russian mathematicians didn't stop here. The *Almagest* contains other resources, such as occultations and lunar eclipses, phenomena that are capable of independent dating. An occultation is similar to a solar eclipse, with the role of the Moon played by a planet, and that of the Sun by a star. Fomenko, Nosovski, and Kalashnikov decided to exploit this information and compare it with the dating of the star catalogue.

The *Almagest* mentions four occultations: Venus with the star η-Virgo, Mars with β-Scorpio, Jupiter with δ-Cancer, and Saturn with γ-Virgo. Ptolemy described these events in detail, relating them to the calendars in use. For example, he presented the occultation of δ-Cancer as follows:

> We again took one of the precisely recorded ancient observations, for which it is said that in the year 45 of Dionysius, on Parthenon 10, the planet Jupiter covered the southern Aselli at dawn. Now the moment is at dawn, in the 83rd year from the death of Alexander, Epiphi 17/18 in the Egyptian calendar.

The word *cover* must be understood in a looser sense than it would be for eclipses. In general, the naked eye cannot distinguish between two celestial objects less than 1 arc minute apart. Some people with very sharp eyes can separate stars as close as 30 seconds, but they are rare exceptions. Since the precision of Ptolemy's observations was no better than 10 minutes of arc, however, his records must be allowed a reasonable margin of error.

Ptolemy dated the four occultations relative to three time-reckoning systems named after ancient rulers: the eras of Nabonassar, Dionysius, and "after the death of Alexander," as shown in table 5.1. But the numbers don't add up. The difference between the occultations of Mars and Jupiter is forty-one years in the era of Alexander and thirty-two in that of Dionysius. This calculation yields two Nabonassar dates: 517 and 508. Fomenko took both possibilities into consideration.

Table 5.1—Dating of the Occultations in the Almagest

Planet involved in the occultation	Years in the era of Nabonassar	Years in the era of Dionysius	Years after the death of Alexander"
VENUS	406	—	—
MARS	476	13	42
JUPITER	—	45	83
SATURN	519	—	—

The numbers in table 5.1 come from the 1952 translation of the *Almagest* by R. Catesby Taliaferro. However, the 1984 version by Gerald Toomer dates the Mars occultation fifty-two (not forty-two) years after the death of Alexander. This revision eliminates the problem: 45 – 13 = 32 and 83 – 52 = 31, so Taliaferro's edition probably contains a typographical error. The discrepancy between thirty-one and thirty-two is not significant because different eras have different starting dates (for instance, the era of Nabonassar begins on February 26, 747 BC, the traditional date of the king's coronation relative to the Julian calendar, and Alexander the Great died on June 13, 323 BC).

The correct numbers would put the occultation by Venus in 341 BC, Mars in 271 BC, Jupiter in 230 BC, and Saturn in 228 BC. But no such sequence of occultations took place in those years or in any neighbouring years. Therefore, traditional chronology deemed Ptolemy's counting of eras unreliable. Using the months (not the years), the longitudes of the Sun and the planets, and the time of the day recorded in the *Almagest*, historians offered the following dates: Venus, October 12, 272 BC; Mars, January 16 or 18, 272 BC; Jupiter, September 4, 241 BC; and Saturn, March 1, 229 BC.

But this solution doesn't fit the physical descriptions in the *Almagest.* Ptolemy wrote that Venus "occulted" η-Virgo, Mars "seemed to have occulted" β-Scorpio, and Jupiter "covered" δ-Cancer. For the traditional solution, the distance between Venus and η-Virgo was 15 minutes of arc (almost half the Moon's diameter); between Mars and β-Scorpio, 50 minutes on January 18 and 15 minutes on January 16; and between Jupiter and δ-Cancer, more than 25 minutes of arc. In each case the angular distance was too large to account for an occultation.

Fomenko didn't object to the identification of the stars, but he disagreed with the dating of the events. The interpretation of Ptolemy's names of the months was doubtful, and the computations of the ancient astronomer had errors similar to those for stars. The Russians chose to ignore the months and the days and consider only the years, the time of the day, and the physical description of the occultations. They found two solutions. The better one had a four-year precision, which was within the margin of error they expected from Ptolemy. It can be described as follows (all hours are relative to London, England):

1 On AD September 9, 887, at midnight, Venus came within 1 arc minute of η-Virgo.
2 On AD January 27, 959, at 6:50 A.M., the angular distance between Mars and β-Scorpio did not exceed 3 arc minutes.
3 On AD August 13, 994, at 5:15 A.M., Jupiter and δ-Cancer were separated by no more than 20 arc minutes, this value being close to the absolute minimum of the angular distance between the two celestial bodies in the time interval from 500 BC to AD 1600.
4 On AD September 30, 1009, at 4:50 A.M., Saturn was at an angular distance of 50 arc minutes below γ-Virgo.

For the last occultation, the *Almagest* mentions that Saturn passed 2 digits below the star. The translations of both Taliaferro and Toomer used the word *digit* for the Greek *dactilos* (δακτιλοζ). Ptolemy mentioned this unit several times when referring to observations made by Hipparchus, who seems to have borrowed it from Babylonian astronomy, where 1 digit is 5 arc minutes. So for the year 1009, the disagreement between Ptolemy's record and the computations is about 40 minutes.

The second solution departs from Ptolemy's conditions, ranging from 329 to 229 BC. Better than the traditional one in terms of matching the years and the physical description, the first solution is not far from the *Almagest* and lies within the margins of error Fomenko expected. Moreover, it agrees with the new dating of the star catalogue. The acceptance of this finding would shift the beginning of

Nabonassar's era to AD 480–90 and shorten ancient and medieval history by more than twelve centuries.

PTOLEMY'S ECLIPSES

There was a third independent verification to do, and Fomenko, Nosovski, and Kalashnikov were eager to learn what this approach would offer. The *Almagest* describes twenty-one lunar eclipses in an interval of 855 years, from 26 to 881 in the era of Nabonassar, traditionally fixed to 747 BC. Eighteen of the descriptions provide the magnitude and relative time span, while three give brief information. The idea was to date the eighteen-event sequence, determine the new era of Nabonassar, and see whether it came close to the AD 480–90 range obtained from the occultations.

Because Ptolemy expressed the dates of the eclipses in different eras starting at different dates, the Russian mathematicians allowed a natural error of one to two years between eclipses and restricted the search to the interval 900 BC to AD 1600. This focus led them to a unique solution, with the first eclipse occurring in AD 491 and the last in AD 1350, setting the era of Nabonassar to AD 465, very close to the previous estimate. But they didn't stop here.

In their view, Ptolemy had made computational errors, sometimes as large as ten years. So they decided to look more carefully at the *Almagest.* Ptolemy had probably taken some of his descriptions from previous astronomers and computed others. His eighteen eclipses can be grouped in terms of eras: the first to the third relative to the era of Mardokempad; the fourth and the fifth, to Nabonassar; the sixth and the seventh, to Darius; and so on, ending with the fifteenth to the eighteenth to the era of Hadrian. The Russians expected that errors occurring within a group were smaller than errors between groups—a reasonable assumption.

They extended the search to allow deviations of up to three years within each group and up to thirty years between groups. The computer showed their previous solution to be still the only one. But allowing an error of four years between any two consecutive eclipses led them to a new solution, in which the first eclipse occurred in 721 BC. This date corresponds to the Nabonassar era starting in 747 BC, in full accord with tradition.

The second solution, however, is worse than the first not only in terms of absolute errors but also relative to how inter-eclipse times are distributed. In the first case the distribution is more even, suggesting consistency, whereas in the second it's more random. This contrast is no proof that the first solution is correct and the second is wrong, but it makes the first look more likely from a mathematical point of view.

Still, there is a problem with the first solution. Since the last eclipse took place in AD 1350, the *Almagest* must have been written after that date, which is outside the likely interval of the star catalogue. In response to this problem, Fomenko said that most lunar eclipses mentioned in the book had taken place around the tenth and eleventh centuries AD, so probably the later ones were added to subsequent copies of the *Almagest.*

This assumption is not unreasonable. Many works deemed ancient, such as the Bible or the Book of Popes, were written by several authors who lived centuries apart. But would historians of science agree with this point relative to the *Almagest*? Perhaps an analysis of the language could reveal whether the prose style changes or help decide about the homogeneity of the composition. Until then, Fomenko's authorship assumption remains conjectural.

What would the implications be of this new dating? The *Almagest* mentions several ancient rulers, so it can help determine when they lived. By placing the Nabonassar era sometime between AD 460 and 490 instead of around 747 BC, Fomenko provided the following dates, all in the Common Era, with an accuracy of plus or minus five years.

The rule of Darius: AD 685–715
The rule of Philadelphus: 840–85
The start of the Callippic periods: 875–910
The death of Alexander: 885–915
The beginning of the Chaldean era: 900–935
The beginning of the era of Dionysius: 915–45
The rule of Augustus: 1175–1205
The rule of Domitian: 1290–1320
The rule of Trajan: 1310–40
The rule of Hadrian: 1310–45
The rule of Antoninus Pius: 1330–65

This chronology shifts the traditional dates forward in time by more than a millennium. It may solve some problems, but it certainly raises others. For example, it qualifies the *Almagest*'s reference to "Alexander." Traditional history has taken this name to refer to Alexander the Great, but it could refer to someone else. There is one emperor, Alexander II of Byzantium, who ruled from AD 886 to 913, a death date falling within Fomenko's range. But Fomenko doesn't explain how this assumption affects the dates of rulers connected to Alexander II.

Among the problems this chronology raises is the connection between Antoninus Pius and the *Almagest.* In his text, Ptolemy used the first year of Antoninus' reign as the epoch of his star catalogue—and for this reason historians think the book was written during the rule of the emperor. Could it be that the original catalogue was later modified in terms of this epoch? It is hard to exclude this possibility, but if many facts were changed or added to the *Almagest*, how reliable is this document for chronological purposes?

Beyond that issue, Fomenko, Nosovski, and Kalashnikov have come up with some interesting methods that can extract new information from the *Almagest.* Like any other historical dating, theirs is based on certain assumptions, different from the ones traditionalists have made. Without forcing the data, and taking three independent approaches that use valid mathematical methods, they have obtained consistent results that have been published in respected journals.

Neither historians nor the scientists who rely on traditional chronology have reacted to these findings. Fomenko's results are obviously not perfect and do not explain everything, but their integrity cannot be denied. Are they sufficient to bring down the flag of tradition? Perhaps not; still, there are other areas in which chronology may prove to be a house of cards.

CHAPTER 6

Ancient Kingdoms

And I will harden Pharaoh's heart, and multiply my signs and my wonders in the land of Egypt.

EXODUS 7:3

Most contemporary scientists dismiss astrology as wishful thinking. But few of them realize that horoscopes are helpful in determining chronology. A horoscope is at the same time both much simpler and much more complex than daily newspapers would lead you to believe.

At its most basic level, a horoscope merely depicts the positions of the Sun and Moon and the planets Mercury, Venus, Mars, Jupiter, and Saturn among the constellations at a given time. Then the issue becomes more complicated. Except for Mercury and Venus, which are never far from the Sun, the other celestial bodies may show up anywhere. As a result, there are 3,732,480 possible configurations of these heavenly bodies.

Because of the planets' fast motion, horoscopes change

almost daily. They may repeat themselves after hundreds or thousands of years or within a few decades. Though finding the date of a particular horoscope involves tedious calculations, a computer can do them within seconds, making the problem easy to solve.

In the 1990s Anatoli Fomenko decided to check several Egyptian horoscopes. He assumed they portrayed either the date of their creation or an earlier date, because monuments or carved inscriptions commemorate the present or the past, never the future. So, any date more recent than the one provided by tradition would support his chronological system—a reasonable assumption. Where does it lead?

A TRIP TO EGYPT

To Gleb Nosovski, it was a dream come true. For years, he had deciphered zodiacs and horoscopes, but only from published drawings. Now, in June 2002, he would finally see the original works. The producers of the Russian TV series *Unknown Planet* had invited him to join their film-shooting expedition to Egypt.

Nosovski had wanted to take this trip years earlier, but he couldn't afford it. The Russian government, depleted of financial resources, had offered few grants for basic research. Fortunately, Fomenko had convinced the Russian media that the story would interest the public, and this project was a result of his team's collaboration with science journalists and filmmakers.

Nosovski was, however, nervous. What if the bas-reliefs and murals he wanted to see were not identical with the drawings he had studied? The relative positions and particular features of signs and symbols were crucial for his results, and the published copies might have misrepresented some elements.

To confirm his theories, Nosovski visited numerous Egyptian sites, including the tombs of several pharaohs. This experience made a deep impression on him. Most of all, he was interested in the carved and painted zodiacs found in those places, so he instructed the photographer to take pictures from different angles and to catch every detail.

The photographs confirmed Nosovski's fears: the previously studied drawings did contain mistakes—but, fortuitously, the originals suited Fomenko's theory better. The majority of horoscopes showed sky configurations that could not have been seen before the Middle Ages, more than a millennium later than historians believed possible. If these horoscopes were more recent, how did Nosovski's conclusions affect Egyptology?

THE DENDERAH STONES

Years earlier, Fomenko had asked Nosovski and other collaborators to investigate Nicolai Morozov's work on horoscopes and, if possible, to develop it. Some Egyptologists had dismissed those carvings and paintings as mere works of art that bore no resemblance to the sky configurations of antiquity. The astronomers' attempts to draw conclusions in agreement with history had failed. But Morozov persisted. In 1930 he wrote about two zodiacs inscribed on the long and the round Denderah stones, which had been discovered in a temple of the goddess Hathor in Upper Egypt, not far from Luxor: "If this were only the fantasy of an artist, then it would be hard to explain why on both zodiacs Mercury and Venus are located near the Sun, as they should be, and not in some place more convenient to the artist . . . No, it's not a fantasy, but a horoscope."

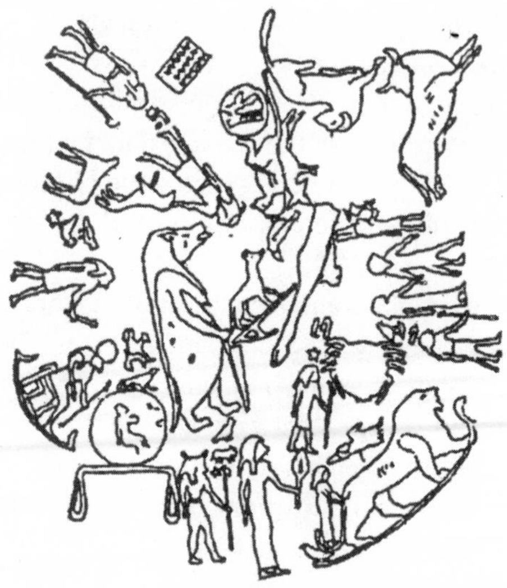

Figure 6.1—The inner part of the round Denderah zodiac—now at the Louvre Museum in Paris—discovered in 1799 by General Louis Desaix during Napoleon's campaign to Egypt. Constellations such as Taurus (the bull), Leo (the lion), and Gemini (the twins) are easy to recognize. The travellers with walking sticks represent planets, in agreement with the Greek word planetes *(πλανητης), which means wanderer.*

Fomenko agreed. The main figures of the round bas-relief are the twelve zodiacal constellations: Aries, Taurus, Gemini, Cancer, Leo, Virgo, Libra, Scorpio, Sagittarius, Capricorn, Aquarius, and Pisces (see figure 6.1). The figures spread among them, birds and travellers with walking sticks, must have represented planets, as was common in medieval astronomical charts. Very likely, the artist was acquainted with astronomy. The problem with ancient representations, however, is to understand the meaning of the symbols. Mixing up two planets may lead to a mistaken date.

The first person to decipher the round Denderah zodiac was the German Egyptologist Karl Heinrich Brugsch. In 1883 he interpreted the figures with the help of the inscriptions found near similar symbols on bas-reliefs, murals, and sarcophagi. Several astronomers tried to date this zodiac using Brugsch's decoding, but their conclusions were dismissed for the simple reason that the years they offered contradicted tradition.

In 1977 the French Egyptologist Sylvie Cauville published a new study on the bas-reliefs of Denderah. In a chapter on the dating of the round zodiac, she used historical reasoning to determine that the temple had been built between 51 and 43 BC, and then she sought astronomical arguments to fix the year. Unable to find a date that agreed with the position of all the planets, she focused on Mercury and Mars. For the former, she proposed June 16, 50 BC, and for the latter, August 12 of the same year. Since then, 50 BC has been accepted as the traditional date for the building of the temple.

To arrive at her initial assessment, Cauville proceeded as any historian would do. She linked the building of the temple with certain archaeological evidence, which she placed between 51 and 43 BC. In other words, she started from conventional chronology and discarded any information not conforming to it.

But, as Fomenko remarked, this approach conflicted with the position of the planets depicted in the Denderah zodiacs. Mars and Mercury move fast, and between June 16 and August 12 they had travelled through several constellations. Also, such relative positions of Mars and Mercury can be found in any century. From the viewpoint of astronomy, therefore, only a horoscope that depicted the positions of all the planets would provide a meaningful date.

Such a solution existed long before Cauville began her study. Morozov had published it in the first volume of *Christ*—a book outside the bibliographical realm of Egyptology. He was interested in dating Egyptian zodiacs and performed detailed computations for several of them. Not surprisingly, he started with Brugsch's decoding of the symbols. Though he disagreed with some of the German expert's conclusions, he based his results on Brugsch.

Morozov had arrived at the dates of AD March 15, 568, for the round zodiac and AD May 6, 540, for the long one. The closeness of the dates favoured the idea that the stones belonged to the same generation. Had they been separated by centuries, Morozov might have doubted the decoding or the chronological significance of the horoscopes. But this outcome led him to believe that the Denderah temple had been built in the sixth century AD.

Morozov missed a few details, however. The figure representing Venus in the long zodiac has the planet appearing in Aries, close to Taurus, whereas Morozov's configuration has Venus at the other end of Aries, near Pisces. This positioning is a tricky matter. If the Egyptian artist was accurate, then Morozov made a mistake. A similar problem occurs with the zodiac's placement of Mercury, west of the Sun between Aries and Taurus; Morozov's date puts it to the east, between Taurus and Gemini.

These aspects raise questions about the hard-to-assess relationship between exactness and artistic freedom, but there is also an inconsistency in Morozov's argument. He interpreted the lack of a star above Mercury as meaning that the planet must have been behind the Sun. But, in the 1990s, the Russian physicists N.S. Kellin and D.V. Denisenko proved that on AD May 6, 540, Mercury was visible (not behind the Sun). The round zodiac poses the reverse problem.

Mercury's figure has a star over its head, whereas in Morozov's scenario the planet is hidden. This enigma prompted Kellin and Denisenko to seek a new solution, in which the round zodiac yielded AD March 22, 1422, and the long one, AD May 12, 1394.

This discovery attracted the attention of Tatiana Fomenko, Anatoli's wife and a research mathematician herself. In 1999 she took a close look at Morozov's sources and learned that he had relied on a drawing of the long zodiac published in 1802 by the French baron Dominique Vivan Denon, who had accompanied Napoleon Bonaparte on his Egyptian campaign. But comparing this drawing with the corresponding picture in *Description de l'Egypte*—a collection of thousands of illustrations commissioned by the French ruler—she noticed a few differences, which, though not huge, could affect the dating of the horoscope.

The main problem concerns a figure with a walking stick, interpreted as Saturn in Denon's drawing (see figure 6.2). Napoleon's album shows no stick at all and, as Nosovski confirmed, neither does the original carving. Tatiana Fomenko redid all the calculations and obtained AD March 15, 568, or AD March 22, 1422, for the round zodiac and AD April 7–8, 1727, for the long one. But she wasn't happy with these dates; they were too far apart to make any sense.

Figure 6.2—Details from the Napoleon (upper) and the Denon (lower) drawings of the long zodiac. The Napoleonic image lacks the walking stick held by the second figure from the right in Denon's copy.

This result prompted several people in Anatoli Fomenko's group to look for additional information. A check of the Egyptological literature confirmed their decoding, but they encountered a problem: the long zodiac depicts four disks, any of which could represent the Moon and the Sun, so the Russian mathematicians took into account all possibilities.

When Nosovski pointed out that certain symbols had no interpretation, they realized they must have overlooked something. That was the case with other zodiacs too. The left-out figures were diverse: a man with a child, a rectangular plate with a wavy line, a self-entwined snake, a two-headed wave-shaped reptile, and so on. A common feature of these figures was their positioning in the same four constellations, which, Nosovski remarked, are those where equinoxes or solstices occur.

From this realization, it was just one step to the idea of a partial horoscope—meant to provide additional information. The smaller figures resembling large characters must be planets indicating sky configurations at solstices and equinoxes. For example, since a falcon-headed figure was identified as Mars, the little character with similar features must have been Mars in the new context.

The partial horoscopes failed to show all the planets, but at least two planets were always present. This information proved enough to find the correct Sun and Moon positions and to confirm the rest of the decoding. Had the partial-horoscope concept been wrong, the team would have found impossible solutions, with equinoxes and solstices not matching the zodiac's date.

A computer did the rest of the work. Out of the many possible dates, only one survived the calculations. The horoscopes for the long zodiac indicated AD April 22–26, 1168, and those for the round one, AD March 20, 1185. Since only seventeen years separated these dates, the team was satisfied with the outcome. Their methodology appeared to be sound, and several attempts to check it were successful.

The Moon gave particularly nice feedback. For the long zodiac, the date of AD April 23, 1168, coincided with the full moon that followed the Easter Moon. Both moons occurred in Libra, and the horoscope contained two lunar symbols. For the round zodiac, AD March 20, 1185, was the Easter Moon day, and the only lunar disk present showed up at the right place.

After so many attempts, Fomenko's team finally had the correct solution. Or so they thought.

THE ESNA BAS-RELIEFS

Using the Denderah stones to date Egyptian zodiacs was the beginning. Fomenko's team decided to study other cases. There was no shortage of them, and the Russian mathematicians focused on the horoscopes from Esna, a town about 50 kilometres south of Luxor. Tatiana Fomenko had already attempted to date the big zodiac from the Khnum Temple and a small one from a nearby building, for which she found the solutions of AD May 1–2, 1641, and AD May 2–3, 1570.

After Nosovski's return from Egypt, the team analyzed his photographs. They soon realized that some essential elements of the horoscopes didn't match the drawings Tatiana Fomenko had studied. The symbol of a planet, for example, occurred in a different place. One such mistake would have been enough to change the result by centuries. But before redoing the calculations, the team checked the decoding time and again.

Figure 6.3—A detail of the big Esna zodiac in which Gemini, Taurus, and Aries appear in the middle. The disk with a crescent is the Moon, and the other disk, the Sun.

The Esna figures are not identical to those at Denderah. For instance, Gemini, which stands next to Taurus, is represented by three figures at Esna and two at Denderah. The first is a man goading a small animal, and the other two are women with folded arms (figure 6.3). The only problem is that the man with the stick might be a planet. But the team rejected this possibility because the stick has no handle, as all those held by travellers do (see figures 6.1 and 6.6).

Since only two disks show up in each case, there were no problems with identifying the Sun and the Moon. For the big zodiac, the Moon is in Taurus (the bull) and the Sun in Aries (the goat, see figure 6.3); for the small zodiac, it's the other way around.

Though the decoding of planets for the big Esna zodiac posed some difficulties, the configuration was a lucky one. Saturn, represented as in Denderah by a male traveller with a crescent on his head, appears at the border of Libra and Virgo, whereas Mercury, Venus, Mars, and Jupiter are all squeezed between Aquarius and Pisces. It is not clear which figure is Mars and which is Jupiter, but the closeness of all these planets is rare enough to make dating easy. A computer analysis showed that only one such period existed in history, from AD March 31 to April 3, 1394.

The ordering of the planets clarified the question of which symbols stood for Mars and Jupiter, a fact the team used for the small Esna horoscope, which was dated at AD May 6–8, 1404. This timing was a pleasant surprise. Given the zodiacs' location in different temples, the Russian mathematicians didn't expect to obtain dates only ten years apart.

Like the Denderah reliefs, the Esna zodiacs contain partial horoscopes, which helped the team test the results. In each zodiac, the equinoxes and solstices fall on the correct days. Though the big Esna stone has no additional scenes, the small one marks the Epiphany—a Christian celebration on January 6 commemorating the arrival of the Three Wise Men to pay homage to Jesus in Bethlehem. On Epiphany day of the year 1404, Saturn, Venus, Mercury, and Jupiter were in Capricorn, as was correctly depicted in a partial horoscope.

This discovery raises the question of why Epiphany, a Christian celebration, would be commemorated by a zodiac. Marking solstices or equinoxes would have made much more sense. But at least this representation is consistent with the fact that, before becoming a Muslim state, Egypt was mostly Christian. The objection that Egypt had been Muslim since the eighth century AD would be valid only if the traditional chronology were assumed to be correct.

One difference between the Esna and the Denderah stones is the meaning of the stars above the heads of planetary figures. Whereas at Denderah the presence of the star means that the planet was visible, at Esna it indicates the opposite. According to Anatoli Fomenko, this change in artistic representation took place during the more than two centuries that elapsed between the construction of the temples.

THE PAINTINGS AT ATHRIBIS

In 1901 the renowned Egyptologist Sir Flinders Petrie discovered two zodiacs at the archaeological site of Athribis, near Suhaj in Upper Egypt. Joined in one composition, they are painted on the ceiling of a burial cave. The division between them is clear and they can be viewed separately, as figures 6.4 and 6.5 show, but there is no doubt that the two zodiacs are the work of the same artist.

The first person to attempt a dating of the zodiacs was the English astronomer Edward Knobel in 1908. His decoding put the upper zodiac in AD 52 and the lower one in AD 59. But Knobel was not happy with his result:

> The year 59 AD, January, suits well for Moon, Mars, Jupiter and Saturn, but is discordant for Venus. No attempt has been made to reconcile Mercury. Jupiter and Saturn would be in similar relative positions about every

Figure 6.4—The upper Athribis zodiac found in 1901 by Sir Flinders Petrie.

Figure 6.5—The lower Athribis zodiac.

> 58 or 59 years. In the epochs 118 BC, 60 BC, 1 BC, 59 AD, 117 AD, the only year that suits the three superior planets is 59 AD, but the position of Venus is quite wrong . . .

This passage leaves the impression that, instead of computing all possible years to which the horoscope might apply, Knobel tried to fit in the ones that suited traditional chronology. Two decades later, Morozov attempted to finish Knobel's work. He looked at all the possible solutions, but none of the dates he obtained fell in the time interval of history. So he changed the interpretation of the planetary symbols, as shown in figure 6.6.

Since the murals were assumed to be the work of the same artist, Morozov imposed the condition that the horoscopes should be no more than thirty years apart. That limitation led him to ascribe the date of AD May 6, 1049, for the upper painting and AD February 9, 1065, for the lower one. But this solution isn't perfect because, on May 6, 1049, Mars

Figure 6.6—Morozov's and Knobel's interpretations of planetary figures for the upper (above) and lower (below) Athribis zodiacs. For each symbol, the upper line shows Morozov's choice, and the lower line, Knobel's choice.

was not at but merely close to the indicated spot. Also, the arguments Morozov used to justify interchanging the symbols for Jupiter and Saturn are not convincing to Egyptologists.

Removing the thirty-year restriction from the original decoding, Fomenko and his collaborators obtained the dates of AD May 15–16, 1230, and AD February 9–10, 1268. In their view, these years might have marked someone's birth and death, and the pictures would have been painted in AD 1268 or soon thereafter. As in previous cases, the solutions accorded with the additional data the team had identified and interpreted in the partial horoscopes, which concerned solstices, equinoxes, and the Easter Moon.

For instance, the summer solstice in the lower zodiac is indicated with the help of the Sun, shown as a man with a raised hand, and five planets, marked by birds, two at the left and three at the right of the man (see the lower part of figure 6.5). The birds look at Gemini, so probably the Sun and the planets were in or near this constellation. The birds at the left have hieroglyphic inscriptions, deciphered by Brugsch as saying Meri-Gor and Ab-Ne-Mani, but the team didn't comment on this aspect.

The left bird in the group of two has the face of a woman, and the team decoded it as Venus, because that is the only female planet. The other bird represents Mercury, and the three birds at the right are Jupiter, Saturn, and Mars. At the summer solstice preceding February 1268, the Sun, Mercury, and Saturn were in Gemini, Venus was in Taurus, Jupiter in Cancer, and Mars in Leo. The Moon, which doesn't show up in the partial horoscope, was in Capricorn, almost opposite to the Sun.

The other partial horoscopes employ similar elements. Unfortunately, the decoding is difficult to justify. Its only strong point is the matching of the dates to provide a meaningful arrangement of the chronology. A broader context would offer even more coherence.

OTHER DECODINGS

Fomenko and his team considered four more horoscopes: one depicted on a sarcophagus discovered by the Egyptologist Heinrich Brugsch in 1856, another painted on a wall in the Petosiris tomb of Dakhla (an oasis some 560 kilometres east of Luxor), and two murals found in the burial chambers of the pharaohs Rameses VI and VII. But, for various reasons, the dates obtained for these paintings are either irrelevant or unconvincing.

The analysis done for the Brugsch horoscope yields a nineteenth-century date, which adds nothing to Fomenko's chronological system. The origin of the Petosiris mural is difficult to establish because the configurations of the planets suggest three possible solutions: in the thirteenth, the fifteenth, and the early eighteenth centuries. Finally, the dates for the pharaohs Rameses VI and VII are inconsistent; according to Fomenko, the latter preceded the former by at least one century. This claim makes no sense, however, because the historical evidence proves that Rameses VII ascended the throne after Rameses VI.

The questionable decoding of the symbols also confirms that the dates obtained for these horoscopes are highly suspicious. So, in the end, only the zodiacs at Denderah, Esna, and Athribis may be relevant for chronological purposes.

OPPOSING EVIDENCE

From this study of Egyptian art, Fomenko concluded that the dates arrived at by his team support his shift theory. Their conclusion is that the history of Egypt is *much* shorter than anyone else has acknowledged and that, as a result, Egyptian culture flourished between the eleventh and the fourteenth centuries AD.

On what facts do these claims rest? Any mathematician or astronomer can verify the accuracy of the team's computations. But what about the interpretation of the symbols and Fomenko's faith in the literal truth of these zodiacs? Could the depicted horoscopes simply have been artistic flights of fancy?

The horoscopes that have been decoded by Egyptologists offer more confidence; still, even those readings are not absolutely certain. Morozov, and then Fomenko, added various elements to the interpretations, not all of which are consistent. For instance, the star above a planetary symbol is viewed ambiguously: at Denderah this depiction means that the planet was visible, and at Esna that it was not.

Whenever partial horoscopes appear, they seem to support Fomenko's claims. But did he make legitimate choices? He often ignored some figures and focused on others. If the creators of those zodiacs intended to represent a date, carefully placing the planets in the right constellations, why would they have added confusing symbols?

The positions of the Sun, the Moon, and the planets are essential information for the correct dating of a horoscope. As mentioned earlier, if even one celestial body is misplaced, the reading of the date can change by hundreds of years. Therefore, any method for decoding horoscopes must be rigorously justified.

Fortunately for the modern chronologist, others have dedicated a great deal of effort to dating zodiacs. In 1959 Otto Neugebauer and H.B. Van Hosen published a study of some two hundred horoscopes, most of them Greek but also a few Egyptian and Arabic. All those horoscopes are explicit, not symbolic, so whenever the text is complete, the interpretation is certain.

Unlike Fomenko and his team, Neugebauer and Van Hosen didn't take every horoscope seriously. They found a few impossible configurations, such as one in which Venus opposes the Sun, but most are plausible from the astronomical point of view. Whether the horoscopes depict the time at which they were drawn is a question that can be answered only by understanding their historical context.

Neugebauer and Van Hosen restricted their study to an ancient interval, ignoring possible dates closer to our time. Viewed as a group, their results are statistically meaningful: the Greek dates range from 71 BC to AD 621, clustering around AD 100; the Egyptian dates fall between

37 BC and AD 93; and most of the Arabic dates are from around AD 800. Their book provides the necessary information for further investigation. There are many more unstudied horoscopes in the papyrological literature, which comprises tens of thousands of texts.

Even if all of Fomenko's solutions were correct, the number of cases he has studied is too small to justify drawing any conclusions. Historians can easily dismiss them as irrelevant because of the uncertainty surrounding their interpretations. Accepting that the history of Egypt is so short poses new problems, which are much harder to solve than the issue of ignoring a few symbols. One such question would be: Where do all the pharaohs fit in?

This problem extends to rulers of other ancient cultures. Documents containing long lists of dignitaries, such as the archons of Athens and the popes in Rome, exist in many archives and museums. If history is a millennium shorter than we think, then the numbers related to the reigns of these men don't add up. But Fomenko thinks they do, and that a mathematical method he developed in the 1980s proves historians wrong.

CHAPTER 7

Overlapping Dynasties

Time present and time past
Are both perhaps present in time future,
And time future contained in time past.

T.S. ELIOT

The idea that history has been "stretched" by the accidental duplication of the records of certain people's lives is eccentric but plausible. If Fomenko had, in fact, identified a dozen or so parallel dynastic pairs in the distant past, then he had a strong case for his dating system. He considered these results the backbone of his theory.

In 1981 the Russian mathematical giant Andrei Nikolaevich Kolmogorov became interested in Fomenko's chronology. After attending one of Fomenko's talks at a conference in Vilnius, Lithuania, Kolmogorov left without saying a word. Later, however, he invited Fomenko to his Moscow residence and asked him to bring along some papers on the subject. Fomenko arrived with a recent 100-page survey and a longer monograph. Two weeks later,

Kolmogorov invited him over again. He asked many questions, most of them on dynastic parallelism. The aging mathematician had sensed the heart of the matter. Fomenko described the conclusion of their meeting by noting that Kolmogorov "said he was frightened by the possibility of a radical reconstruction of a number of modern concepts based on ancient history. He had no objection to the legitimacy of the methods. Finally . . . [he] asked whether he could keep the 100-page essay as a present."

This reminiscence failed to explain whether Kolmogorov had commented on Fomenko's methodology or whether he had been silent about it. In any event, it was clear that Kolmogorov saw some merit in Fomenko's investigation. The idea of parallelism had been in the air before.

GRAPHS AND DISTANCE

In his *Chronology of Ancient Kingdoms Amended*, Isaac Newton argued that Europeans had made certain parallel kings successive and sometimes presented one king as two. Nicolai Morozov went further by identifying several dynasties that tradition had viewed as belonging to different periods. But the first scientist to study this issue using mathematical methods was Anatoli Fomenko.

He attached to each dynasty a sequence in which every entry represents a ruler's term. For example, if the first king reigned for twelve years, the second for eighteen, the third for seven, the fourth for twenty-three, and the fifth for nine, the sequence of the dynasty is (12, 18, 7, 23, 9). It can be represented as a graph, with the horizontal axis giving the king's dynastic position (first, second, etc.) and the vertical axis showing the length of reign (figure 7.1).

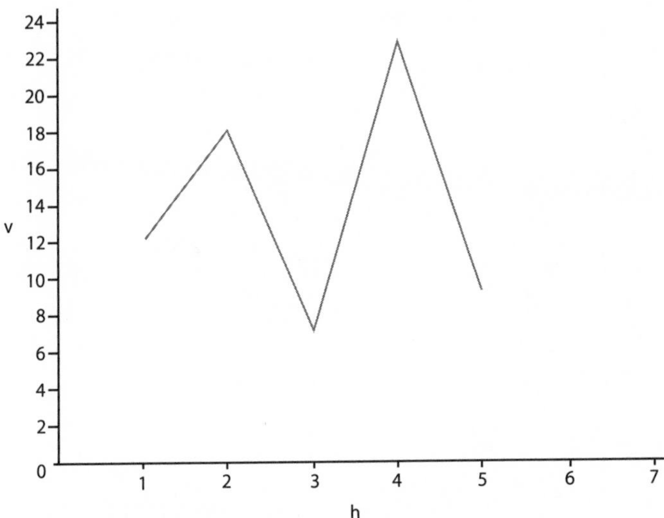

Figure 7.1—The graph of the sequence (12, 18, 7, 23, 9).

Comparing the graphs of two sequences makes it possible to detect any resemblance between them. For instance, the sequence of the Holy Roman–German Empire (AD 911–1376) is almost the same as that for the Kingdom of Judah (931–586 BC) (see figure 7.2).

However, the graph alone is a rough tool. To distinguish between different degrees of closeness, Fomenko defined a distance in the set of finite sequences. In common language, distance is the extent of separation between two points. On a flat surface, the distance is measured along the straight line that connects them. But if the points are, say, Toronto and Rome, you must estimate the length of an arc of a big circle on a sphere that approximates the Earth. In other words, distances depend on the environment.

Mathematics has found ways to define distances on any surface of any shape or dimension. This idea can be

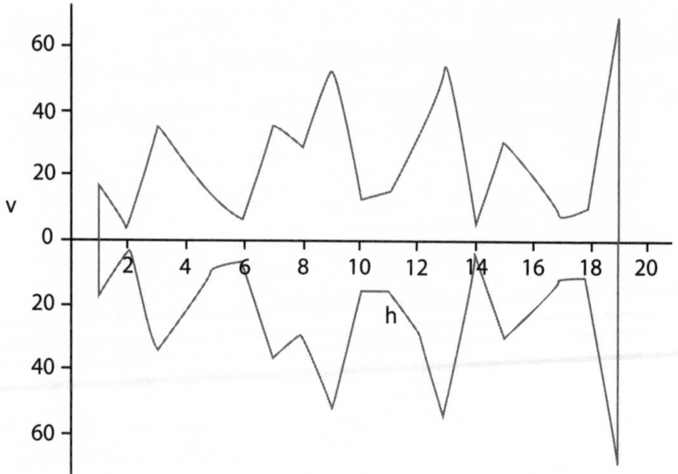

Figure 7.2—A comparison between the dynastic graphs for the Holy Roman–German Empire (AD 911–1376), upper graph, and the Kingdom of Judah (931–586 BC), lower graph. For convenience, the graphs are drawn symmetrically.

extended to abstract spaces, which may have no intuitive geometrical representation, such as the set of finite sequences. Those distances are by no means arbitrary or ambiguous; they satisfy precise properties and provide reliable tools in estimating the "closeness" of any two objects.

Fomenko introduced such a distance to measure the likelihood that dynasties from different epochs were identical. After comparing thousands of dynastic sequences, he found thirteen pairs that seemed to overlap:

1 Carolingian Empire (AD 6–9th centuries) and Third Roman Empire (AD 3rd–4th centuries)

2 Holy Roman Empire (AD 10–13th centuries) and Third Roman Empire (AD 4–6th centuries)
3 Holy Roman–German Empire (AD 10–13th centuries) and Imperial House of Hapsburg (AD 13–17th centuries)
4 Holy Roman–German Empire in Italy (AD 10–13th centuries) and Second Roman Empire (1st century BC–AD 3rd century)
5 Holy Roman–German Empire (AD 10–14th centuries) and Biblical Kingdom of Judah (10–6th centuries BC)
6 Roman coronations of the Holy Roman Emperors (AD 10–13th centuries) and Biblical Kingdom of Israel (10th–7th centuries BC)
7 First Roman Pontificate (AD 141–314) and Second Roman Pontificate (AD 314–532)
8 First Roman Empire (regal Rome, 753–500 BC) and Third Roman Empire (AD 3rd–4th centuries)
9 Second Roman Empire (82 BC–AD 3rd century) and Third Roman Empire (AD 3rd–6th centuries)
10 Biblical Kingdom of Judah (capital in Jerusalem, 10–7th centuries BC) and Eastern Roman Empire (capital in Constantinople, AD 306–700)
11 Biblical Kingdom of Israel (10th–8th centuries BC) and Third Roman Empire (AD 4th–5th centuries)
12 First Byzantine Empire (AD 527–829) and Second Byzantine Empire (AD 829–1204)
13 Second Byzantine Empire (AD 867–1143) and Third Byzantine Empire (AD 1204–1453)

Using the same method, Fomenko compared the events of ancient and medieval Greece, 10th to 3rd centuries BC and AD 10th to 16th centuries, and noticed that their sequences were very similar. He finally found statistical duplicates within different periods of the Trojan War. In his view, these results indicate that time frames with similar chronologies are identical, and therefore history is much shorter than has been assumed.

IDENTICAL POPES

Particular examples hold the key to understanding the details of Fomenko's method. The early papacy (item 7 in the above list), which in his opinion forms "the spinal column of medieval Roman history," is a suitable subject of inquiry. The Russian mathematician presented his argument as set out in table 7.1: he compared the years in office for two sequences of nineteen popes; the number at the end of each entry indicates the pope's length of rule in years.

Table 7.1—Comparative Sequencing for Nineteen Popes

First Roman Pontificate (AD 141–314)	Second Roman Pontificate (AD 314–532)
1 Pius I (141–57), 16	1 Silvester I (314–36), 22
2 Anicetus (157–68), 11	2 Julius I (336–53), 17
3 Soter (168–77), 9	3 Liberius I (352–67), 15
4 Eleutherius (177–92), 15	4 Damasus I (367–85), 18
5 Victor I (192–201), 9	5 Siricius (385–98), 13
6 Zephyrinus (201–19), 18	6 Anastasius I Innocent (398–417), 19
7 Calixtus I (219–24), 5	7 Boniface I (418–23), 5
8 Urban I (224–31), 7	8 Celestine I (423–32), 9
9 Pontianus (231–36), 5	9 Sixtus III (432–40), 8
10 Fabian (236–51), 15	10 Leo I (440–61), 21
11 uncertain (251–59), 8	11 uncertain, Hilarius (461–67), 6

12 Dionysius (259–71), 12	12 Simplicius (467–83), 16
13 Felix I (or Eutychianus?) (275–84), 9	13 Felix III (483–92), 9
14 Eutychianus (or Felix I?) (271–75), 4	14 Gelasius (492–96), 4
15 Gaius (283–96), 13	15 Symmachus (498–514), 16
16 Marcellinus (296–304), 8	16 Hormisdas (514–23), 9
17 Marcellus I (304–9), 5	17 John I (523–26), 3
18 Eusebius (309–12), 3	18 Felix IV (526–30), 4
19 Meltiades (311–14), 3	19 Boniface II (530–32), 2

The graphs in figure 7.3 suggest the numerical resemblance of the two groups. Indeed, Fomenko used the distance defined in the set of finite sequences as well as some statistical methods to show that the two lists of pontiffs can be identified.

Fomenko argued that A and B (A being the sequence of years in office for the popes of the first Roman pontificate and B representing the same for the second pontificate) showed interesting similarities:

$$A = (16, 11, 9, 15, 9, 18, 5, 7, 5, 15, 8, 12, 9, 4, 13, 8, 5, 3, 3),$$
$$B = (22, 17, 15, 18, 13, 19, 5, 9, 8, 21, 6, 16, 9, 4, 16, 9, 3, 4, 2).$$

Sequence A and sequence B are "very close" when compared with other sequences in the same set. Taking into consideration the errors in recording and transmitting data over time and ignoring all other information about those popes (such as name, place of birth, relatives), the resemblance of A and B might indicate identical historical periods. Thus Pius I and Silvester I might represent the same person, Anicetus might be the same as Julius I, and so on in the preceding list.

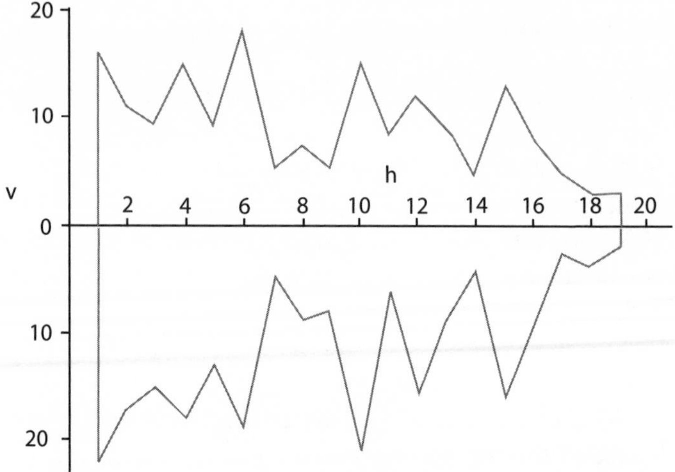

Figure 7.3—The graph of Fomenko's comparison in table 7.1 between the popes of the first and the second Roman pontificates.

Since "near" and "far" are relative notions, this "closeness" makes sense only if most other sequences are far from each other. Fomenko's computations showed that most dynastic distances were large. Though he used statistical estimates to obtain his result, he knew that he possessed no rigorous proof for his method. However, in the context of probability theory, he deemed this outcome very likely and therefore suitable for historical purposes.

In particular, Fomenko emphasized the nonexistence of close dynasties from the sixteenth century to the present, a conclusion supporting the accuracy of the traditional interpretations for the past five hundred years. This recent period of history is the only one that Fomenko never questioned.

But if two periods coincide, he concluded, the dynasties must be identical, parallel, or a combination of both. So, either homologue kings are the same person or they ruled simultaneously. In the latter case, you would expect to see evidence of interaction between them. Fomenko can therefore be shown to be mistaken if identification and interaction prove illusory.

Theologians and historians alike have recognized the importance of the papacy as the only European institution connecting the present to the ancient past. Consequently, many researchers have focused on understanding its social role and, over several centuries, a wealth of historical material pertaining to the papacy has accumulated.

A key document about the early popes was written by Eusebius of Caesarea (*c.* 260–341), one of the bishops present at the Nicaean Council (see chapter 2). Other popes are included in *Liber Pontificalis* (The Book of Popes), a series of manuscripts compiled by several clerics beginning in the fourth century AD. These works often contradict each other, and researchers are still struggling to compile an accurate history of the first pontiffs. Nevertheless, theologians and historians agree upon a significant amount of information. Without it, neither the early history of the papacy nor Fomenko's identification would make any sense.

Fomenko noticed that the data about the first pope, St. Peter, and his seven successors, until Pope Hyginus (*c.* 140), was uncertain and unsuitable for use. He began his study with Pius I, who was pope for sixteen years, from 141 to 157. Though vague in its references to the election and termination dates of the first popes, *Liber Pontificalis* is often precise about the length of their terms in office. But historians rarely take this information for granted; different sources provide different numbers. Fomenko gave a term of sixteen years for Pius I and twenty-two for Silvester I; other references mention fifteen and twenty-one, respectively.

These variations do not, however, demolish Fomenko's argument. The two sequences would stay close were it not for the occurrence of other popes, whom Fomenko omitted. Table 7.2 compiles a different set of data for the same periods: 140 to 314 and 314 to 532. In this new list, I have included in bold face the popes he left out, except for Anastasius I and Innocent I, whom Fomenko mentioned as one pope.

Table 7.2—Revised Comparative Sequencing for the Popes

Popes from 140 to 314	Popes from 314 to 532
1 Pius I (140–55), 15	1 Silvester I (314–35), 21
2 Anicetus (155–66), 11	2 **Marcus** (Jan.–Oct. 336), 9 months
3 Soter (166–74), 8	3 Julius I (337–52), 15
4 Eleutherius (175–89), 14	4 Liberius I (352–66), 14
5 Victor I (189–99), 10	5 Damasus I (366–83), 17
6 Zephyrinus (199–217), 18	6 Siricius (385–98), 13
7 Calixtus I (217–22), 5	7 **Anastasius I** (399–401), 2
8 Urban I (222–30), 8	8 **Innocent I** (402–17), 15
9 Pontianus (230–35), 5	9 **Zosimus** (Mar.–Dec. 418), 9 months
10 **Anterus** (235–36), 44 days	10 Boniface I (418–22), 4
11 Fabian (236–50), 14	11 Celestine I (422–32), 10
12 **Cornelius** (251–53), 2	12 Sixtus III (432–40), 8
13 **Lucius I** (253–54), 8 months	13 Leo I (440–61), 21
14 **Stephen I** (254–57), 3	14 Hilarius (461–68), 7
15 **Sixtus II** (257–58), 1	15 Simplicius (468–83), 15
16 Dionysius (260–68), 8	16 Felix III (483–92), 9
17 Felix I (269–74), 5	17 Gelasius I (492–96), 4

18	Eutychianus (275–83), 8	18	**Anastasius II** (496–98), 2
19	Gaius (283–96), 13	19	Symmachus (498–514), 16
20	Marcellinus (296–304), 8	20	Hormisdas (514–23), 9
21	Marcellus I (307–9), 2	21	John I (523–26), 3
22	Eusebius (Apr.–Sept. 309), 5 months	22	Felix IV (526–30), 4
23	Meltiades (310–14), 4	23	Boniface II (530–32), 2

A notable difference in this list is the growth of the number of popes in each pontificate from nineteen to twenty-three. Had one column included more entries than the other, the distance between sequences would have been large, and the identification could have been dismissed. But sequences C and D, where

$$C = (15, 11, 8, 14, 10, 18, 5, 8, 5, 1/10, 14, 2, 2/3, 3, 1, 8, 5, 8, 13, 8, 2, 5/12, 4) \text{ and}$$
$$D = (21, 3/4, 15, 14, 17, 13, 2, 15, 3/4, 4, 10, 8, 21, 7, 15, 9, 4, 2, 16, 9, 3, 4, 2),$$

have the same number of elements, so it is necessary to compute their distance (see also figure 7.4). My computations showed that C and D are far apart, which seems to contradict the overlapping of the Roman periods from 140 to 314 and from 314 to 532. But before drawing any conclusion about the overlaps, a closer analysis of the new list turns out to be useful.

A first occurrence in the second column is that of Marcus, who was pope for about nine months, from January to October 336. Since he was the successor of Silvester I, the entire second pontificate is shifted one entry down, invalidating the subsequent identifications. The other new

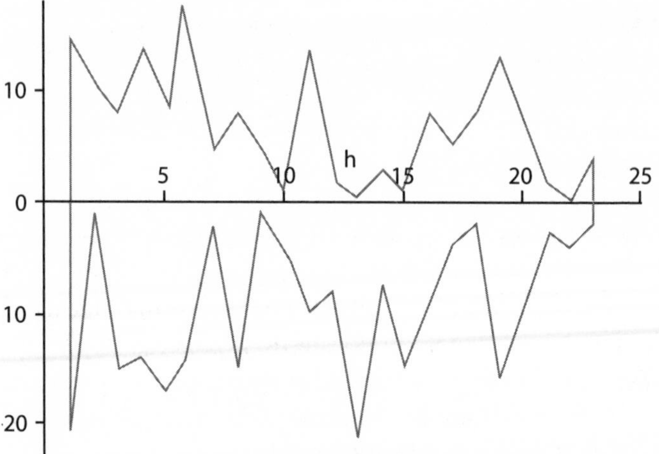

Figure 7.4—The graphs of the first and second Roman pontificates, as given in the modified list in table 7.2.

pontiffs—Anterus, Cornelius, Lucius I, Stephen I, and Sixtus II in the first column, and Anastasius I, Innocent I, Zosimus, and Anastasius II in the second—restore Fomenko's comparison, but only for his last five entries.

However, the new popes in the second list had either very short terms of office or appeared in periods of confusion, which explains why Fomenko excluded them. Therefore, it isn't obvious that this list invalidates his conclusion. This debatable issue could hardly provide a straight answer, so I decided to regard the original problem from a different angle.

Assuming Fomenko's list was accurate, I wanted to check other numbers—for example, the length of the periods when some popes coexisted with antipopes (pontiffs who were supported by powerful cardinals who, in turn, opposed the official elections). Often, this duality was possible because of a voting system that proved easy to exploit in the context of the political instability of those times.

The first antipope mentioned in history is Hippolitus (217–*c.* 235), who overlapped with Calixtus I, Urban I, and Pontianus (first column of table 7.1, Fomenko's selection). The corresponding popes in the second column are Boniface I, Celestine I, and Sixtus III. There is indeed an antipope during the time of Boniface I—Eulalius (418–19)—but he was in office for a much shorter time than Hippolitus, so it is unlikely that these two antipopes represent the same person. A second antipope is Novatian (251–*c.* 258), who appears in a period of confusion and has no homologue between 461 and 467.

Relative to the second column of Fomenko's list, the first antipope is Felix II (355–65), a contemporary of Pope Liberius I, but he has no match in the first column during the time of Soter. A similar case occurs with antipope Ursinus (366–67) at the end of Liberius I's office and the beginning of Damasus I's; he finds no correspondent in the first column at the time of transition between Soter and Eleutherius. Antipopes Laurentius (498 and 501–5) and Dioscorus (530) have no counterparts in the First Roman Pontificate either.

Still, most antipopes from 140 to 532 appear for short intervals of time or during periods that are not clearly documented, so, again, it is difficult to claim that their presence is a clear refutation of Fomenko's theory. Like the short-term popes, they may have been included in some documents and overlooked in others.

A convincing argument for or against Fomenko's identification of the two Roman pontificates is difficult to make without investigating the popes' lives more thoroughly. For example, consider Meltiades (311–14) and Boniface II

(530–32), who occupy the last position (entry 19) in Fomenko's list.

Meltiades is also known in the literature as Miltiades, Milciades, Melciades, Melchiades, and Miltides. His year of birth is unknown, but he seems to have been of African origin. He became pope on July 2, 310 or 311, died on January 10 or 11, 314, and was buried in the Catacomb St. Callistus. After the death of his predecessor, Eusebius, who had held the office for about five months, the Roman See was vacant for at least four and at most seventeen months. During Meltiades's time, Constantine the Great entered Rome after his victory over Emperor Maxentius at Milvian Bridge on October 27, 312. Soon after that, Constantine presented the Roman Church with the Lateran Palace, which became the seat of the pontifical administration.

On a close look, Meltiades's homologue looked quite different. Boniface II was elected pope on September 17, 530, and died on October 17, 532. His tomb is in St. Peter's, Rome, where a fragment of his epitaph has been preserved. He was the first pontiff of German origin, son of the Ostrogoth Sigisbald. During the reign of his predecessor, Pope Felix IV, Boniface served as archdeacon, having a considerable influence with the ecclesiastic and civil authorities.

Fearing death and a subsequent contest for the papacy between Roman and Gothic factions, Felix IV had gathered his clergy and several Roman senators and solemnly conferred on his aging archdeacon the papal sovereignty, proclaiming him his successor. He threatened anyone opposing him with excommunication. On Felix's death, Boniface assumed the succession, but sixty of about seventy cardinals refused to endorse him and instead elected Dioscorus as antipope. They feared the influence in papal affairs of the Ostrogothic king Athalaric, who was close to Boniface. Both popes were consecrated on September 22, 530, Boniface in the Basilica of Julius, and Dioscorus in the Lateran. The crisis lasted only twenty-two days, for Dioscorus died on October 14.

As history has presented them, the life sketches of Meltiades and Boniface II are far from similar. It is improbable that crucial events, such as the succession crisis during the time of Boniface II, have been overlooked in Meltiades's case. It is hard to believe that a pope of African origin elected in July (summer), dead in January (winter), and buried in the Catacomb St. Callistus was the same person as a pope with German roots who assumed office in September (fall), died in October (fall), and was laid to rest in St. Peter's. The two pontiffs also differ in the reforms they made, the politics they promoted, and their overall attitude towards believers. On these facts, any objective commentator would find it hard to identify Meltiades and Boniface II as one and the same.

A biographical analysis of other popes leaves a similar impression. Apart from the apparent closeness between sequences, no other data support Fomenko's claim. His critics understandably want him to clarify why the lives of homologue pontiffs look so dissimilar.

It is tempting to dismiss the overlapping theory right away, but the question remains: Why would Fomenko insist on repeating what seemed like an obvious mistake with more than a dozen pairs of dynasties? This question becomes even more intriguing when one considers that Fomenko has provided yet another challenge, some might say a provocation: the coincidence of Carolingian kings and some Roman emperors.

CAROLINGIAN KINGS AND ROMAN EMPERORS

The dynastic identification of emperors and kings is quite different from the one involving popes because it usually deals with the succession to the throne of blood descendants or other relatives. The first overlap Fomenko noted is the one between the Carolingian kings (681–888) and the emperors of the Third Roman Empire (324–527). His comparison is given in table 7.3.

Table 7.3—Coincidence of Carolingian Kings and Some Roman Emperors

Carolingians: Charlemagne's Empire in the 6th–9th centuries	The Third Roman Empire in the 3rd–6th centuries
1 Pépin of Héristal (681–714), 33	1 Constantinus II (324–61), 37
2 Charles Martel (721–41), 20	2 Theodosius I (379–95), 16
3 Pépin the Short (754–68), 14	3 Arcadius (395–408), 13
4 Charlemagne (768–814), 46	4 Theodosius II (408–50), 42
5 Carloman (768–71 or 772), 3 or 4	5 Constantin III (407–11), 4
Charlemagne's donation of Italian lands, 774	Donation of Constantine of Rome, 4th century
6 Louis I the Pious (814–33), 19	6 Leo I (457–74), 17
7 Lothar the Western (840–55), 15	7 Zeno (474–91), 17
8 Charles the Bald (840–75), 35	8 Theodoric (493–526), 33
9 Louis the German (843–75), 32	9 Anastasius (491–518), 27

10 Louis II the Western (855–75), 20	10 Odoacer (476–93), 17
11 Charles the Fat (880–88), 8 Dissolution of the Carolingian Empire in the West. Shift by *c.* 360 years. War.	11 Justin I (518–27), 9 Dissolution of the official Third Roman Empire in the West. Gothic War of the 6th century AD.

The sequences E and F, in which

$$E = (33, 20, 14, 46, 3 \text{ or } 4, 19, 15, 35, 32, 20, 8) \text{ and}$$
$$F = (37, 16, 13, 42, 4, 17, 17, 33, 27, 17, 9),$$

result from this list, and are indeed close to each other (see figure 7.5). However, this situation is more complicated than the one concerning the popes. Here Fomenko doesn't give a complete list of rulers of a certain country, but selects a variety of kings and emperors who, according to tradition, reigned over different territories.

Theodosius I and Charles Martel offer a characteristic example. Both have resonant names in history: the former, the last emperor to rule briefly over a united Roman Empire (392–95) and the one who imposed Orthodox Christianity as a state religion in the East; the latter, renowned for his victory in the Battle of Tours-Poitiers (October 732), which has been hailed as the pivotal event in Europe's stand against the Arab invasion.

Theodosius I was born at Cauca in Galicia (northwestern Spain) around 346, the son of Theodosius the Elder. In 375 he became governor of Moesia and in 378 co-augustus for the East, appointed by the emperor Gratianus. From 392 onward he ruled alone. In 394 he defeated the armies of the usurper Eugenius at the Battle of Frigidus.

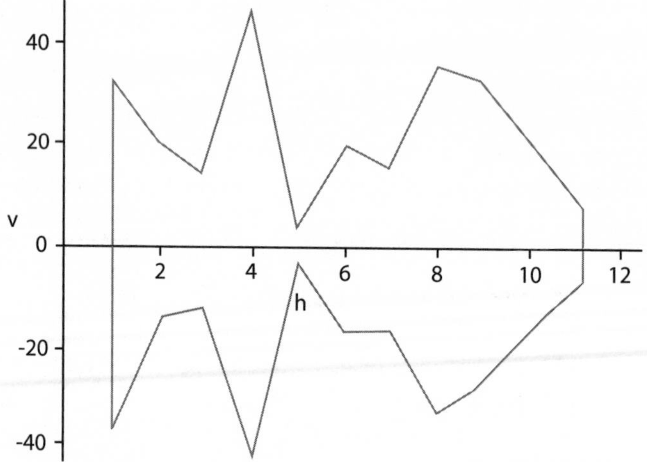

Figure 7.5—The dynastic graphs of the Carolingian and the Third Roman Empires.

Theodosius was married twice. His first wife gave him a daughter and two sons, one of whom, Arcadius, succeeded him to the throne. By his second wife he had a daughter. He died of an illness in Milan in January 395.

In contrast, Charles Martel was the illegitimate son of Pépin of Héristal, a mayor of the palace for a Frankish king. Between 638 and 751 the Frankish kings were rulers only in name, spending their lives in amusement and leaving state affairs in the hands of their mayors. At the death of Pépin in 714, Charles fought his stepmother and other rivals to become mayor of the palace, succeeding around 721. Apart from defeating the Muslim invaders, he led the Frankish army against several states to the south and east of his kingdom.

He had two wives. One of the two sons from his first marriage, Pépin the Short, later became king. Charles

Martel died in October 741 at Quierzy in today's Aisne department of the Picardy region in France.

As in the case of the popes Meltiades and Boniface II, Theodosius I and Charles Martel seem to represent very different individuals. Furthermore, history makes no reference to any contact between them, so the idea that they might have been contemporary appears to have no basis. But how are things with other homologue rulers in the list? The most important historical personality here is Charlemagne, whom Fomenko has identified with Theodosius II.

In 408, when he became Roman emperor of the East, Theodosius II was seven or eight years old. The power rested with the praetorian prefect Anthemius, who had served Theodosius' father, Arcadius. In 414 Theodosius' sister Pulcheria was made regent and led the empire until her brother came of age.

Theodosius fought two wars against the Persians, in 421–22 and in 441–42, both of which he won. But he was unsuccessful in his conflicts with the Huns, who had settled in the Pannonian plains. Between 441 and 443 the Huns raided the Balkan provinces and defeated the imperial armies, asking for a heavy tribute in exchange for staying away from the region. In 447 they attacked again, and Theodosius gave in to all their demands. This capitulation prompted King Attila of the Huns to regard himself as the overlord of the emperor.

Among Theodosius' administrative achievements was the Codex Theodosianus, a collection of Roman laws imposed in 438 and accepted in both the East and the West. The emperor also established the University of Constantinople. In 450 he died unexpectedly in a riding accident, failing to leave a successor.

Though not a negligible figure, Theodosius can hardly

be compared to Charlemagne, also known as Charles the Great, king of the Franks and the Lombards, founder of a new Roman Empire in the West, unifier of civilizations with the blessing of the Catholic Church, and one of the most important rulers in history. Europe regards Charlemagne as an inspiring symbol in the context of its present economic union.

Born in 742, he became king, jointly with his brother Carloman, at the age of twenty-six. The two co-rulers didn't get along, risking a division of the kingdom. But Carloman died in 771 or 772 of natural causes, and Charlemagne assumed power alone. By 810, his territory included central and western Europe to the North Sea and the Pannonian plains, half of the Italian peninsula, Corsica, and the north-eastern part of Spain, south of the Pyrenees.

Unlike Theodosius II, Charlemagne was at war throughout most of his reign. The culminating moment of his life occurred on Christmas Day in the year 800, when Pope Leo III crowned him emperor of the Romans during a ceremony that filled St. Peter's Basilica with enthusiastic supporters. The coronation strengthened his alliance with the Catholic Church.

While Theodosius II paid tribute to the Huns, Charlemagne fought their close relatives, the Avars, for many years. His victory was so decisive that the Pannonian plains looked deserted at the end of the war. This rout prompted the Russians to coin the phrase "vanishing like the Avars," a saying they use to this day.

Charlemagne's cultural achievements were also impressive. He was a patron of the arts and sciences, inviting to his capital, Aachen, learned people from many fields: astronomers, architects, musicians, and grammarians. For good reason, this period is known as the "Carolingian

Renaissance." The emperor himself wrote a grammar of the German language spoken in those days and worked on improving the translation of the Gospels. This cultural spirit lasted for a century after his death; western Europe never again receded into the darkness that had loomed before his time.

At the end of January 814, depressed by the recent deaths of his wife and his sons Pépin and Charles, the aging emperor developed a high fever with pleurisy and died after a few days. He left the throne to his son, Louis of Aquitaine, who later became known as Louis the Pious. Charlemagne was buried in an Aachen church that still exists today. The marble sarcophagus in which his body was allegedly laid to rest is preserved in the church's treasury.

Fomenko's identification of Charlemagne with Theodosius II seems as unlikely as all the others. Moreover, the choice of rulers in the list is not always comprehensible. For instance, while the inclusion of Louis I the Pious is correct, because he followed Charlemagne, his homologue Leo I did not come to power immediately after Theodosius II. Marcian reigned before him, between 450 and 457. He is not the only skewed choice in the list. Leo II, for example, who ruled together with Zeno for ten months, is omitted.

This second example of parallelism appears even less convincing than the first. Not only do the biographies of the rulers not resemble each other, but even the dynastic sequences are questionable. Not surprisingly, Fomenko has an answer to this problem too.

THE STABILITY OF THE METHOD

Fomenko regards ancient and medieval history as a puzzle in which many pieces are missing. Most documents that have survived from this period are vague and incomplete, and therefore open to interpretation. To solve the puzzle, historians need a backbone, a basic structure, which is provided by the landmarks of chronology. Once these generally accepted landmarks are in place, historians can arrange the existing pieces and fill in the missing ones with more or less good guesses. In time, however, these guesses become established as truths, making it difficult to distinguish between the backbone pieces and the conjectural ones. Every new piece of evidence is then interpreted in relation to what is already known; the historical edifice depends more and more on the initial arrangement.

Fomenko thinks that the biographies of the rulers discussed above are unreliable because they have been compiled by interpreting many vague and incomplete documents in the light of an erroneous chronology. He believes that if Scaliger and Petavius had offered different landmarks, many of these rulers would have vanished from history, and the biographies of the surviving ones would have looked different.

Assume, for instance, that the biography of a ruler has been compiled from three documents, *A*, *B*, and *C*. In *A* he is referred to as *x*, in *B* as *y*, and in *C* as *z*. The accepted chronology and some apparently connecting event suggest that he is one and the same person, called *x*, y, or *z*. But under a different chronology, the connection fails to occur, and the character becomes a dubious figure if not an entirely fictitious one.

This scenario is difficult to prove or disprove. Fomenko would have to experiment with a new chronology and reconstruct all of recorded history from scratch, redoing the four hundred years of work since Scaliger and Petavius. Moving any piece of the puzzle would affect most of the pieces around it. Such a titanic enterprise cannot be finalized by a handful of people; it would require a huge multidisciplinary effort, one that might be impossible to complete because some of the documents used earlier have not survived.

The whole issue becomes a dilemma. On the one hand, historians cannot invalidate Fomenko's conclusions with historical data alone because he can always identify in their reasoning a circular argument based on traditional chronology; inevitably, all their attempts to refute him are flawed. On the other hand, Fomenko is unable to convince the experts because he offers too little evidence.

Fortunately, there is a way out of this conundrum. The validity of the dynastic method can be clarified from a purely mathematical point of view, without any reference to history.

The main question concerns the stability of the method. This notion has tens of connotations in mathematics, but here it means that small perturbations of the initial data lead to nothing beyond small changes in the outcome. In other words, if there is a slight variation in the information provided for the dynastic lists, the distance between dynasties changes very little. A violation of this condition means that the method is unstable, and therefore not very useful.

Fomenko is well aware of this problem, as any mathematician would be. He has spared no effort to prove that if the rulers' lengths of reign vary a little, the distance between dynasties doesn't change much. His result is correct, but insufficient. He has brushed aside an important aspect of the problem: the entries on his lists may also change. In other words, he may have omitted some rulers or added others at the wrong place, as can be seen by comparing his list in table 7.1 with my list in table 7.2. In those examples, the changes have led to large variations of the distance, as figures 7.3 and 7.4 show.

This detail shows that the dynastic method is unstable at larger perturbations and is therefore unreliable as long as the entries in Fomenko's lists are uncertain—which, unfortunately, is and will remain the case in ancient and medieval history. Consequently, the overlaps he has indicated are only apparent, and an entry's addition or omission is prone to disprove the identifications. So the main pillar of his revised chronology turns out to be very shaky.

Fomenko has, however, designed other methods with applications in historical research. His language analysis of texts, comparisons of ancient and medieval maps, and etymological investigations provide radical critiques of and alternatives to traditional historical approaches to the past.

CHAPTER 8

Secrets and Lies

There are three kinds of lies:
lies, damn lies and statistics.

BENJAMIN DISRAELI

We all assume that mathematics is essential to the physical sciences, such as chemistry and physics. We are prepared to admit that it may be useful in the social sciences, including psychology and anthropology. So, what about the humanities? Can mathematics play any useful role in a field that tends to look askance at the quantitative aspects of reality? Specifically, can statistical methods contribute anything to the study of history? Anatoli Fomenko would certainly answer in the affirmative.

EMPIRICAL-STATISTICAL METHODS

Many people are suspicious of statistics, and not without reason. Numbers have often helped sustain distorted truths.

But mathematical statistics is not responsible for its misuse. As a branch of mathematics, it can extract the beneficial information that is often hidden behind mountains of data. It helps distinguish the forest from the trees.

When chronology first started to interest him, Fomenko was not an expert in statistics. But he soon realized there was more to be read in historical texts than the stories they told. He had only to use the right mathematical tools and a wealth of new knowledge would surface. After teaching himself the statistical theory he would require, in the late 1970s and early 1980s he developed a few dating techniques.

One of his methods was aimed at identifying various chronicles that appear to be different but describe the same historical period, even if they appear in different languages, use different geographic names, and give their characters different names. Indeed, a person, deity, country, or city can be known by more than one name: George W. Bush is also referred to as Dubya, God as Allah, Finland as Suomi, New York City as the Big Apple. Poor communication may have allowed name variations to abound in the past. It is reasonable, then, to think that such chronicles exist.

The method works as follows. Consider two texts describing several historical events that have a relative dating but not an absolute one. Start by collecting various data: the number of words used to chronicle a year or the number of times a name occurs in a decade. For the same type of information in each text, draw graphs and compare them. If the curves are very different, the periods are probably unrelated. If they look similar, continue the investigation.

A particular feature to watch in a pair is the simultaneity of peaks, as in figure 8.1(a). If, say, the graphs represent the number of words that chronicle the reign of a king, then summit points on the curve indicate more words and identify eventful years. If those dates agree, the texts might describe the same period. This process is called the amplitude correlation principle, and pairs of curves that violate it, such as the pair in figure 8.1(c), are discarded. In figures 8.1(a) and (b), the graphs are correlated. The difference between them is that, in the latter graph, the "dotted" chronicle lays more weight on the middle event than the "continuous" one does. Nevertheless, the principle is satisfied.

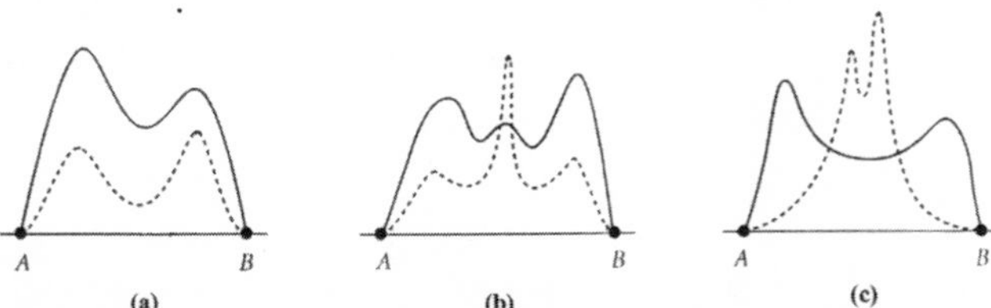

Figure 8.1—Each pair of graphs represents similar types of information collected from two different chronicles. In (a) and (b), the amplitude correlation principle is satisfied, whereas in (c) it is not.

This procedure runs the risk of overlooking texts covering the same event. Consider, for example, a pair of chronicles, where one is written by the winner in a conflict and the other by the loser. The former may dedicate a lot of text to the victory, whereas the latter may barely mention the defeat. But since the goal is to find identical periods, the prospect of missing some cases strengthens, though also narrows, the method.

The graphs are compared with the help of certain mathematical devices that allow a quantitative evaluation of each text. As a result, even the most accessible of Fomenko's books contain mathematical formulas that tend to obscure his ideas and make understanding difficult for the non-expert.

After designing the statistical model and laying out the hypotheses, Fomenko tested his method on reliable historical writings from the sixteenth to the twentieth centuries to find out whether different chronicles known to describe the same events satisfied the amplitude correlation principle. He was pleased to see they did. Then he compared several Russian texts, among them the Suprasl' and the Nikiforov chronicles, both allegedly referring to events occurring in the period AD 850–1256. Their word graphs—which appear in figure 8.2—are remarkably similar. Fomenko also examined documents known to describe distinct periods, and all violated the principle.

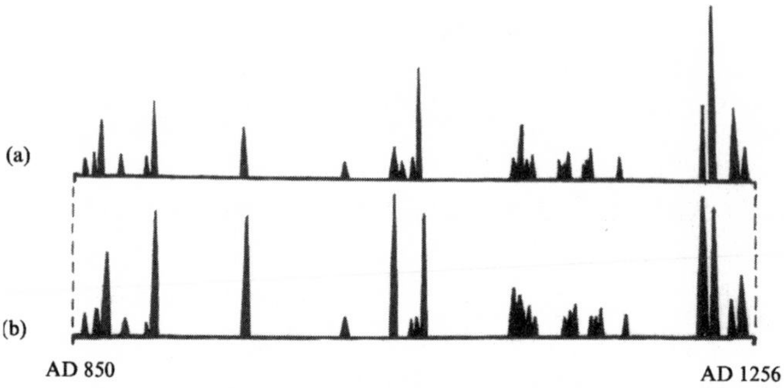

Figure 8.2—The word graphs of the Suprasl' (a) and the Nikiforov (b) chronicles.

Another problem was to order chronologically several writings that contain hundreds of historical characters, some of whom appear in many of the documents. Fomenko thought of the following solution. Divide the texts into chapters, one for each generation span. In any given chapter, only names from the past or the present show up—not those from the future. Then introduce a quantitative measure:

compare the occurrence of names from previous generations with those in the investigated chapter and write down the ratios. Since parents are better remembered than grandparents, more distant generations are less frequently mentioned; therefore the ideal graph of the ratios would look like the one in figure 8.3. The chronicles are then ordered such that all mutual frequencies are close to the ideal ones.

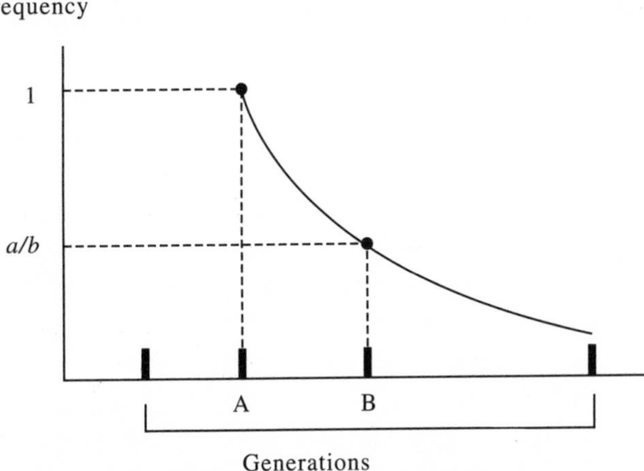

Figure 8.3—The horizontal axis represents all generations. If, in chapter B, generation A is mentioned a times and generation B, b times, the vertical axis indicates the ratio a/b. Before the birth of generation A (to the left of A), generation A is not mentioned, so the ratio is 0. At A, the ratio is a/a = 1. At the time of generation B, generation A is mentioned less frequently the farther B is from A. Therefore the ratio a/b decreases for more distant generations, so the ideal frequency graph looks like the one above.

Fomenko called this hypothesis the frequency damping principle and checked several reliable documents of the last few centuries to see whether they agreed with it. He claimed they all did. Now he had a method to identify texts of the same period and another to order those of distinct ones. Together these principles could help him determine the relative chronology of the chronicles.

THE HOLY BIBLE PROJECT

Fomenko applied his statistical method to several texts. Even with the help of his assistants, this detective work took a few years. Among his investigations, his analysis of the Holy Bible was a significant proving ground. The Bible project was particularly important because it allowed an independent verification of traditional chronology. According to Scaliger, the Scriptures described distinct events, except for the overlap between the four Old Testament books of Samuel and Kings and the two books of the Chronicles.

Fomenko suspected that the Bible was not chronologically ordered. To prove this point, he had to sort the events it described by using the frequency damping principle. But his method was not easy to apply because the Scriptures contain about two thousand names, which are mentioned tens of thousands of times. So he divided the Old and New Testaments into 218 chapters, one for each generation of characters.

The division he arrived at was uneven with respect to the original books. Each time the narrative changed direction, a new chapter began. For instance, Genesis in Fomenko's classification contained seventy-three chapters: Genesis 1–3 (Adam, Eve), Genesis 4:1–4:16 (Cain, Abel), and so on, whereas Exodus formed only one chapter. The Old Testament consisted of parts 1 to 191, and the New Testament, 192 to 218. To test the validity of this division, Fomenko checked the method on the already acknowledged biblical overlaps and located them without difficulty.

Then he shuffled all the generation chapters according to the outcome of his statistical analysis. The conclusion is astonishing: the Old and New Testaments describe some interwoven events and are not separated by several centuries, as Scaliger had thought. For example, the Revelation of St. John the Divine, the last book in the Bible, is assumed to be a part of the New Testament. Placing it anywhere else would look strange at best. But Fomenko's frequency analysis points at a position near the prophecies of the Old Testament. His new ordering moves Revelation to the same period as the books of Isaiah, Jeremiah, Ezekiel, Daniel, Exodus, and Leviticus. Fomenko did not find this placement surprising because St. John's Revelation reminded him of the apocalyptic nature of Daniel's prophecies in the Old Testament.

But this one example does not prove that the events of the Bible must be reordered and reinterpreted; it is only an indicator in this direction. Without a close analysis of the text's meaning, any conclusions are premature. To my knowledge, no biblical scholar has yet investigated this issue in the spirit Fomenko suggested. The method, however, has promise. It sheds some light on new aspects for scholars to investigate, and for that alone it is worthy of attention.

DATING MAPS

Fomenko used a similar method in geography. His approach in this case is based on the following principle, which he tested in 1979 and 1980. Consider a chronologically correct sequence of maps. Then assume that, in experiments, once an error is corrected, it doesn't reappear in subsequent maps and all the accurate features are maintained. For example, if a portion of land depicted as a peninsula proves to be an island, the correction is noted and it will show up as an island on all future maps. Consequently, for a region with a long cartographic history, the fewer the number of mistakes the map contains, the more recent the map.

Given a sequence of maps whose dates of drawing are unknown, one can order them chronologically by comparing every pair and noting down the changes that occur from one to the next. (This process is similar to the frequency analysis for generation chapters that Fomenko used for texts.) Many criteria must be taken into account, including the type of map (globe, flat); the kind of projection (conic, cylindrical, azimuthal); orientation; the arrangements of poles, equator, and tropics; the representation of climatic zones; and so on.

This idea is not new to historians. Sir Flinders Petrie, the father of modern archaeology, used a similar technique at the beginning of the twentieth century after noticing the stylistic differences between the articles of pottery found in various graves. By charting those changes, he determined the graves' relative chronology.

Since Fomenko had few ancient maps at his disposal, the verification of his method was somewhat hindered. Eventually, however, he obtained a classification that passed several tests, including an obvious one: maps with similar styles were drawn in the same period.

According to his findings, cartography developed very slowly. The maps of the third and fourth centuries AD were simple sketches, very different from what they depicted. Then their quality improved, until the occurrence of the first fairly accurate globes and planar maps in the 1500s. But though the Earth's main features were present in those drawings, they still posed serious proportion problems.

Several famous maps attracted the attention of the Russian mathematician. One was attributed to Claudius

Ptolemy in the second century AD. The version Fomenko referred to had appeared in a 1551 edition of Ptolemy's *Geography.* With the help of his method, Fomenko concluded that this chart had the features of maps created in the fifteenth and sixteenth centuries AD. This result was not far from his dating of the *Almagest* (see chapter 5).

A time difference of more than a millennium showed up between the traditional and the new dating of the Tabula Peutingeriana map thought to have originated in the time of Emperor Augustus (27 BC–AD 14). Fomenko found that it had the cartographic characteristics of the eleventh and twelfth centuries. For several other maps, traditionally dated to the eighth to second centuries BC, as well as the globe of Crates (second century BC, see figure 8.4), Fomenko attributed dates between the seventh and the thirteenth century AD.

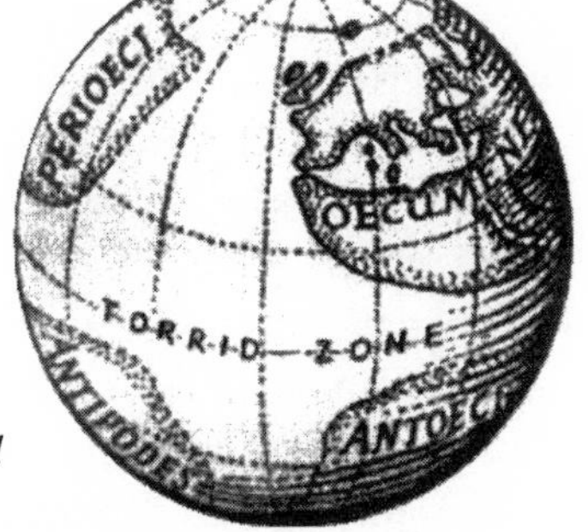

Figure 8.4—Reconstruction of Crates's globe, traditionally dated to the second century BC.

Disregarding the conclusions, the principles of Fomenko's method meet the same standards historians apply when dealing with the evolution of patterns, as Sir Flinders Petrie did in his analysis of pottery. Therefore, as in the case of the empirical statistics for texts, this idea seems worth pursuing. But can these methods withstand criticism on other fronts?

THE CRITICS RESPOND

From the mathematical point of view, there is nothing wrong with Fomenko's techniques; they use the standard tools of mathematical statistics, which nobody questions. But whether his conclusions reflect chronological reality is another issue. The discipline that deals with the interpretation of such data is known as applied statistics, and its expert practitioners are aware of its traps. For example, was Fomenko's pool of data relevant and rich enough? Did any experiment fail? These questions can be answered only by independent verifications. That was difficult to do in the Soviet Union in the 1970s, but it doesn't pose a challenge now, in the age of computers.

Another problem concerns the comparison of the conclusions with the historical reality. The testing does not prove the method correct—it shows only that it works in some cases. But it's not difficult to imagine examples in which it fails. In truth, Fomenko's method of analyzing a historical text can at best find candidates for a solution. It is essential, then, to check which of those potential dates agree with the logical meaning of the document. There is no evidence Fomenko did that. Without this confirmation, the technique can fail, as it would, for example, if it placed a chapter in which a character dies before the chapter in which he is born.

For maps, the considerations are different. Although Fomenko claimed he could give absolute dates, it is doubtful that it can be done without relying on other results. As in the case of texts, he might be able to order charts, but how can he tell if the earliest maps are from the fourth century BC or the seventh century AD? Such conclusions cannot follow from statistical methods without relying on external reference points. Fomenko needs to fix at least one date by different

Figure 8.5—Ptolemy's 1482 Ulm edition of the World Map on a spherical projection.

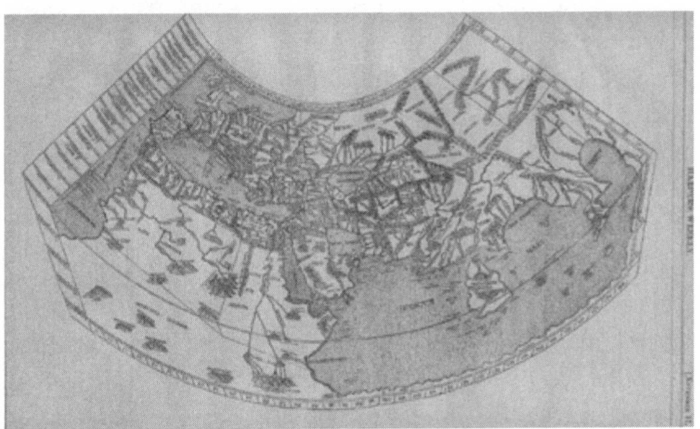

Figure 8.6—A redrawing of Ptolemy's map, published in 1877 in Harper's Weekly.

means; otherwise his relative scale can be shifted, stretched, or contracted.

Against all appearances, dating maps is more difficult than dating texts. If some ancient original writings have survived to this day, even if in the form of medieval copies, their content is generally similar to the original. That is not true for maps. The overwhelming majority of purportedly ancient charts are copies made in the Middle Ages, and they reflect the superior cartographic techniques of that period.

A good example of this problem is Ptolemy's world map. Fomenko dated its 1551 Basileae edition to the fifteenth or sixteenth century. Is this surprising? Ptolemy's map has been redrawn many times in the past six hundred years, and each new imprint has distinct features. A comparison of the 1482 Ulm version (figure 8.5) and the 1877 *Harper's Weekly* edition (figure 8.6) shows these differences. But neither map can reveal much about the original. That is not the case for the *Almagest*, whose astronomical records allow better estimates of its provenance.

The experts' views about Fomenko's empirical-statistical methods are mixed. Some critics, like the probability-theory expert A.N. Shyraev, are enthusiastic about his work. In 1990 Shyraev wrote:

> The scientific results obtained by the author [Fomenko] are most remarkable indeed, and what we witness today can already be referred to as the rather sudden evolvement of a whole new scientific division in applied statistics that is definitely of interest to us. All of the results in question were deduced from a tremendous body of work performed by the author with the assistance of his fellow academicians, most of them specializing in mathematical statistics and its applications.

Others, however, think little of these methods. Among them is Owen Gingerich of the Harvard–Smithsonian Center for Astrophysics, who said: "The whole thing is as nutty as can be, and you can quote me on that." A similar opinion comes from the Russian mathematician Sergei P. Novikov, recipient in 1970 of the Fields Medal—the mathematical equivalent of a Nobel Prize. In a three-page article entitled "Pseudohistory and Pseudomathematics: Fantasy in Your Life," published in *Russian Mathematical Surveys* in 2000, Novikov harshly criticized all aspects of Fomenko's work, from geometry and differential equations to history and chronology, by conveying the oral statements of specialists in various fields. One of those he cited was S.A. Aivazyan, whom Novikov counts among the best Russian experts in applied statistics. Novikov wrote that Aivazyan "had already considered Fomenko's historical-statistical materials and found that they contained nothing at all: the emperor wore no clothes, and there was nothing to discuss."

However, Novikov presented no arguments to back up the allegations. That is unfortunate, especially as Fomenko and Novikov co-authored an excellent geometry book, translated into English in 1984. But as Novikov confessed, their personal relationship went downhill after this collaboration. Half of his article deals with one of his letters, which Fomenko published without his consent. As a result, and unfortunately, Novikov's comments are highly emotional and subjective.

THE TRAPS OF ETYMOLOGY

The same issue of *Russian Mathematical Surveys* includes the article "Linguistics According to A.T. Fomenko," signed by another member of the Russian Academy, A.A. Zalyzniak. If Novikov lacked arguments, Zalyzniak had plenty. Most of them referred to *The New Chronology and Conception of the Ancient History of Russia, England, and Rome*, a book Fomenko and Nosovski had published in Moscow in 1996. No English translation of this work exists, though a short version is posted on the Web.

The two mathematicians tried to prove in *The New Chronology* that the history of England is much shorter than historians believe and that many events thought to have occurred in Britain had actually taken place in the Byzantine Empire. The premise is unorthodox enough, but the methodology of the book is equally eccentric, considering the authors' expertise: no mathematics, no statistics, no astronomy—only linguistics: the origins of words. The authors used etymology, a treacherous field in which an amateur can easily make wrong guesses.

Zalyzniak was in his element and had no difficulty putting the two mathematicians in a bad light. He rarely ventured outside his territory, but when he did he paid the price. One of his slips concerned Thucydides' eclipses. Another revealed his idealized image of dendrochronology, a dating technique discussed in chapter 9. Zalyzniak's linguistic conclusions, however, were very well argued.

In opening, he noted that mathematicians could apply their methods to whatever field they liked, including chronology, but only after solving some preliminary problems regarding the collection of data and the translation of the issue into mathematical terms. When the information is unreliable, as often happens in history, no theorem can lead to a valid result.

Zalyzniak warned that his approach was not aimed at those who believed in "Fomenkology" because it sounded revolutionary or courageous. He wanted to address readers who were ready to listen to arguments. Then he tore into Fomenko and Nosovski's book, whose linguistic level indicated the "most primitive and ignorant dilettantism," and compared its blunders to the mathematical equivalent of making mistakes in the multiplication table.

The authors, Zalyzniak observed, put too much emphasis on consonants, practically ignoring the role of the vowels. According to them, the word *Lithuania* has the backbone *LTN*, and therefore comes from the word *Latin.* And the words *Turks* and *Trojans* have similar roots, *TRK* and *TRN*, and are therefore related. Indeed, it is known that the early writing system of Semitic languages such as Arabic, Hebrew, and Phoenician was based on consonants, but that is not true for Greek, Latin, Russian, or English.

Another frequent mistake Zalyzniak noticed was Fomenko and Nosovski's reversing of words, on the ground that this is the way they are read in some Eastern languages. For example, the word *Samara*, denoting a major city in southeast Russia on the Volga, is, to the two authors, *A-Ramas*, which, having the backbone *RM*, means Rome. The presence of the letter *s* didn't seem to matter.

This error stemmed from the authors' mistaken view that the written language had priority over the oral one. Languages can survive, maintain their basic features, and even develop without a writing system. The Slavonic dialects called Lusatian or Sorbian, for example, existed in oral form in a German environment for more than five centuries. Others, such as the Romany spoken by the Roma, or Gypsies, are still alive today.

Fomenko and Nosovski didn't specify the language or period to which the words they dealt with belonged. They did not merely fail to inform the reader but considered it unimportant. That, however, disregards a basic rule of linguistics. To an amateur, a word is an entity in itself, whereas the linguist sees it as a member of various classes. For instance, *culture* belongs to the class of two-syllable words, the class of words starting with *c*, the class of nouns derived from Latin, and so on.

The authors took the freedom to transcribe every word in simplified Russian, an unobjectionable practice. However, they then used this transcription to draw conclusions that cannot be reached otherwise. A striking example was the biblical name *Rosh*, which, according to Fomenko and Nosovski, the medieval Byzantines simplified to *Ros* because they knew this meant *Rus* (Russian). But Greek, the language in which the text in question was written, has no sound *sh*. Whenever *sh* appeared in a foreign name, the Greeks substituted it for the letter sigma (denoted by Σ or σ), which reads *s*. The Hebrew name *Sholomon* turned into *Solomon* (Σολομων), and the Akkadian *Assur* (s = sh) became *Assyria* (Ασσυρια).

Another blunder, in Zalyzniak's opinion, occurred when Fomenko and Nosovski extended the meaning of a word to different unrelated languages. For instance, they said the Greek *basileus* (βασιλευζ), which means king, made the Russian name *Vasilij* denote the same. They identified Vasilij Blazhennyj (Basil the Blessed), a sixteenth-century miracle man from Moscow, with the first of the four biblical kings, who in their opinion was Ivan the Terrible. The same logic made the Russian word *Turkmeny* (an inhabitant of Turkmenistan) signify Turkish men.

To help the reader understand the magnitude of these mistakes, Zalyzniak explained that a discipline named comparative-historical linguistics had been in existence for more than two centuries. Its goal is to provide methods for distinguishing related words from those that happen to sound alike.

Linguistic changes are neither individual nor random—they follow rules. One is regularity: a sound turns into another not only for some words but for all. However, each modification is limited to a language and to a historical period. Take, for example, the English sound *th* and the German *d*. The English *this*, *then*, *feather*, and *bathe* have the German analogues *dies*, *dann*, *Feder*, and *baden*. Their genetic affinity does not necessarily make them sound alike, but they obey phonetic rules of correspondence.

Linguists distinguish between affinity and borrowing. The latter can take place between related languages, say, English and German, or unrelated ones, such as English and Japanese. An example of borrowing within the first group is the German word *Gesundheit*, also used in English as a good-health wish to a person who has just sneezed. In the second group falls the word *harakiri*, which in Japanese as in English denotes an honourable form of suicide.

When checking if a word *x* of a language *A* is related to a word *y* of a language *B*, linguists first choose between affinity and borrowing. For affinity, they must ascertain whether the rules of phonetic correspondence are obeyed. For borrowing, they must determine the direction, say from *A* to *B*. Then they take each sound of the word *x* and see how it changes in *y* within the context of language *B*.

Another point to establish is whether *y* has suffered additional variations specific to loaned words. When earlier versions of *x* and *y* are known, they have to be compared. If the meanings of *x* and *y* differ, linguists must learn the reason and track down the history of the problem. Once a hypothesis has passed these tests, it is compared with competing hypotheses. Only then can conclusions be drawn.

But Fomenko and Nosovski displayed a complete ignorance of these principles. They identified all similar sounds as the same sounds—in particular, those within the groups s-z-sh-zh, b-v, v-f, f-t, t-d, k-h-g, k-ts-s, g-z-zh, ch-sh-shch, r-1, and n-m. For example, *gus* (goose) is *guz* (Turkish tribe) and *sever* (north) is *Sibir* (Siberia). *Vrag* (enemy), *varyag* (Viking), and *fryag* (Frenchman or Italian) are identical. The same applies for *Shchek* (a man's name) and *chekh* (a Czech), for *ulus* (a region of the Golden Horde) and *Rus* (Russian), and for the Byzantine *kir*, the English *sir*, and the Russian *tsar.* Again, the vowels were ignored.

Of the same type was the identification of the towns *Terebovl* and *Tver*, since the backbones *TRB* and *TRV* are identical according to the rule that *b* equals *v.* But that never happens in Russian. Such a phonetic change did, however, take place in Greek: β was pronounced *b* in antiquity but *v* during the Byzantine period. The authors probably compared English and Russian names like *Barbara* and *Varvara* or nouns like *alphabet* and *alfavit.* These examples exist because Western languages borrowed from Greek through Latin, which retained the ancient pronunciation of β, whereas Russian was under Byzantine influence.

Zalyzniak used a metaphor to explain the linguistic logic of Nosovski and Fomenko. "In arithmetic it would look like this: the square of a number often ends in the same digit as the number itself, doesn't it? 1 x 1 = 1, 5 x 5 = 25, 6 x 6 = 36. Why shouldn't we suppose that 7 x 7 = 47?" This analogy would be the mathematical equivalent of what they did: threw away all the vowels, removed or commuted consonants, and equated one consonant with another within each resemblance group for the words of any language. Taking such liberties, anybody could "prove" anything.

An aside: Fomenko and Nosovski share the tendency of one of the characters in Nia Vardalos's 2002 comedy *My Big Fat Greek Wedding.* There, the father of the bride is obsessed with the idea that every English word comes from the Greek. He sets out to "prove" his point any time the opportunity presents itself. His etymological blunders are very similar to the ones the Russian linguist mocks in his article.

Zalyzniak provided a few more examples to show the difference between an amateurish guess and a linguistic analysis. One of them referred to the Russian word *musulmane* (Muslims), which, according to Fomenko and Nosovski, comes from the city of *Mosul* in Mesopotamia. Though they didn't explain the role of *mane* in *musulmane*, Zalyzniak guessed they meant *man*, as they had when they equated *Turkmeny* with Turkish men.

Traditional etymology has reached a different conclusion. In Arabic the word is *muslim(un)*, the ending *un* being dropped in some cases, and it means "obedient (to God)" or "giving one's safety and soundness (to God)." The root *SLM* is the same as for the words *salam(un)* (peace or safety) and *islam(un)* (Islam, which means obedience). The Arabic name of Mosul, *al-Mawsil(u)* (knot or point of link, from the root *WSL* = to link), contains a different *s* from the one in *muslim(un)*. Russian doesn't distinguish between the two, so they look identical in transcription.

Persian borrowed the word and added the suffix *an*, thus creating the forms *musilman* and *musulman.* Those led to the identical Kazakh and Tatar versions *musulman* and, finally, to the Russian plural *musulmane*, derived according to a rule obeyed by other Russian words such as *gorozhane* (city dwellers), *moldavane* (Moldovians), or *khristiane* (Christians). Its singular form is *musulmanin.*

Zalyzniak couldn't find a single correct interpretation among those that filled Fomenko and Nosovski's book. Everything appears to be guesswork. To him, this theory "bears the same relation to a scientific investigation as a report on the authors' dreams." The response of the two Russian mathematicians was the same as it had been in other cases. Traditional linguistics, like traditional history, is strongly influenced by the existing chronology. The way this science was founded and developed depends on what language was there first and on who influenced whom.

There is some truth in this view, but the statement is wrong in general because many etymological principles are independent of chronology. The direction of borrowing depends not necessarily on history but on the structure of each language. For instance, the Russian word *oktobr* (October) stems from Latin. This connection doesn't follow because the Romans preceded the Slavs, but because Latin has the word *octo* (eight), out of which *October* and many other Latin words were derived. Or take the Russian *zakuska* (appetizer, hors d'oeuvre) and the French *zakouski.* Linguists recognize a Russian prefix, root, and suffix, so here the French have borrowed from the Russian.

Unfortunately, Fomenko and Nosovski compared only those words that sound alike. In fact, it would have been difficult for them to follow the principles of linguistics because an etymology based on the new chronology doesn't exist. To obtain credible results towards further research on historical dating, they must take into consideration only chronologically independent conclusions or build a linguistic theory based on dates established by other means.

Another aspect Zalyzniak addressed in his article was document falsification. This section came as a response to the authors' assumption that, mostly in Russian history but also in that of other countries, many written texts thought to be ancient had been forged in the Middle Ages. Though the Russian linguist had good arguments against this claim too, not all of them were independent of traditional chronology. Nevertheless, he made his point.

Zalyzniak also outlined a principle he saw emerging from Fomenko's work: no records of the past can be trusted. They have been corrupted by forgetfulness, mistakes, or intentional lies. There are, however, exceptions: those that contain information convenient to Fomenko. For instance, he ignored Russian texts that have a date written on them, such as Ostromir's Gospel, 6,565 years "from the creation of the world" (AD 1057), or the Svyatoslav Izborniks collection, inscribed with 6,581 (AD 1073) and 6,584 (AD 1076). But he was sure about the length of the reigns of those rulers who allowed him to point at dynastic parallelisms.

Zalyzniak also commented on the authors' style. The book is written negligently. It jumps from one topic to another and includes many repetitions and misspelled words. Instead of *Holmgard*, they wrote *Holmgrad*; sometimes they referred to *Kenigsberg* and at other times to *Keningsberg*; often *gezhdra* appeared as *gedzhra* and *tamga* as *tagma*, and so on.

On page 45 of their book, Fomenko and Nosovski insist that no Russian parchment chronicle survived. Then on page 391, when mentioning the Laurentian Chronicle of 1377, they state it was written on parchment. "But of course," Zalyzniak sneered, "how are the authors supposed to know what they would write on page 391 when they were on page 45?"

In Zalyzniak's opinion, Fomenko's ideas wouldn't have drawn much attention in Russia had the author not been such a prominent mathematician. The Russian public is impressed with the status of his profession and his membership in the

Russian Academy, and people think that the methods he developed, with their elevated standards of logic and rigour, prove whatever he claims. The reality, however, is different. In his linguistic analysis, Fomenko not only didn't refer to mathematics, the domain in which he is an acknowledged expert, but completely ignored the rules of linguistics, a field about which he apparently knows little.

Fomenko, Zalyzniak continued, tries to impress with his honours and titles, which are emphasized in the introductions to his books. Pompous sentences show up everywhere: "We believe that the unprejudiced reader is already convinced that we are moved by the firm logic of scientific research"; "we have to move further along this path if we want to stay on the ground of common sense and strict scientific principles"; "[our conclusions] follow directly and unambiguously from medieval Russian documents"; or "the identification of Russia in a certain period with Ireland follows unequivocally from old English chronicles."

To Zalyzniak, "this is the position of a prophet, of a guru, of the leader of a religious sect, but not of a scholar." While admitting that history has its problems, he disagreed that Fomenko had solved them. In the concluding section of his article, Zalyzniak wrote:

> Fomenko's doctrine in its present form cannot even play the role of a useful stimulus for prompting serious historians to put the dark corners of traditional chronology in order. This doctrine has already slipped through the stage when it could have a claim to such a role. Having piled mountains of amateurish nonsense and phantasmagoric fabrications on top of chronological problems, ignoring professional science and addressing instead the unprepared public, Fomenko has so firmly placed

> himself outside science, as well as outside common sense, that future investigators of chronology will not bother to dig through this mountain of absurdities to look for a rational kernel buried in it.

Zalyzniak was also concerned that Fomenko "shamelessly" exploited the authority of mathematics and tarnished its good name. Clairvoyants, fortune tellers, and believers in superstition and magic could now arrogantly claim that they do science. Based on Fomenko's example, people who maintained that "South America was initially inhabited by Russians," "Peter the Great was a woman," or "Nicholas II and Leo Trotsky are the same person" would label those who disagreed with them as rigid conservatives. To Zalyzniak, no one has damaged the prestige of mathematics more than Fomenko.

Zalyzniak may seem somewhat unusual to the Western reader. His article is sarcastic in the extreme, often rude and offensive, taunting Fomenko and Nosovski whenever an opportunity occurs. Every page of it brims with anger and scorn. Zalyzniak sees the two mathematicians as arrogant outsiders who ignore the research of other disciplines, look with disdain at the humanities, and think they are the only ones who understand Truth.

Had he not been so open in his animosity towards Fomenko and Nosovski, his article might have been better received. To Zalyzniak, everything Fomenko has done is rubbish, and even if there might be something good in his work, it's hidden behind mountains of nonsense. But an impartial reader cannot refrain from asking why the Russian linguist is so bitter.

Zalyzniak is undoubtedly correct about linguistics. Fomenko and Nosovski have blundered terribly by ignoring the basic rules of a discipline they haven't mastered. Are they compromised? What about their other results? And, when extending his verdict to unfamiliar fields, isn't Zalyzniak guilty of making judgments in areas of which he knows little?

Before one tries to answer these questions, it's useful to understand the scientific dating methods, especially the radiocarbon technique. Are they objective tests of physical evidence? Do they support Fomenko's work, do they come closer to tradition, or do they provide a different perspective?

PART THREE

Science Fights Back

CHAPTER 9

Scientific Dating

Everything that has come down to us from heathendom is wrapped in a thick fog . . . We know that it's older than Christianity, but whether by a couple of years or a couple of centuries, or even by more than a millennium, we can only guess.

RASMUS NYERUP

It was one of the greatest discoveries of the twentieth century. In 1947 a young Bedouin shepherd climbed into a cave at Qumran, some 20 kilometres east of Jerusalem on the shore of the Dead Sea, and found a pottery jar containing several scrolls (figure 9.1). He brought these curiosities out of the cave; eventually, the documents fell into the hands of antiquity dealers, who offered them to scholars.

This discovery triggered a decade-long search of the area, an archaeological enterprise that led to the discovery of more than eight hundred manuscripts, in tens of thousands of pieces, most of them made of animal skin but also some of papyrus and copper. Written in Hebrew and Aramaic, they contain biblical and commercial texts, sectarian books and letters.

Figure 9.1—The archaeological site at Qumran, near Jerusalem, where several of the Dead Sea Scrolls were found.

From the beginning, experts disagreed on when the Dead Sea Scrolls had been made. Some considered them to predate the birth of Jesus; others thought they were of medieval origin. But while archaeologists kept busy digging at the site and rummaging through the caves, news spread about the invention of an independent dating method that could reveal the age of any document. An unexpected breakthrough was about to change the field of chronology.

RADIOCARBON DATING

In January 1948 a group of thirty people, including several archaeologists and anthropologists, attended a talk in New York given by Willard Frank Libby, a former member of the Manhattan Project and now a chemistry professor at the University of Chicago. Libby presented a new theory for dating objects in areas such as archaeology and geology. He asked if anyone could donate pieces of known-age wood or charcoal to help him check his results.

At the end of the talk the room remained silent until Richard Foster Flint, a geologist from Yale University, said he liked the idea and offered to send samples. Then other people began to ask questions, showing their interest in the method. This lecture was the first public disclosure of a scientific achievement that would make repeated headlines in the years to come.

In 1951 Libby tested fragments of the linen in which some Dead Sea Scrolls had been wrapped. The results agreed with what the majority of archaeologists expected: the cloth had been made sometime between the second century BC and the second century AD. This early success convinced Libby to focus on refining the technique, whose margins of error were still too large. In 1956 and 1960 a team of experts checked palm-wood samples found at the Qumran archaeological site and estimated them to have been cut between 15 BC and AD 10, plus or minus eighty-five years.

This revolutionary invention made Libby world-famous, and in December 1960 he was awarded the Nobel Prize for chemistry. But his joy was short-lived. More and more archaeologists began to express concerns about the accuracy of his dates, and some even began to doubt the age of the Dead Sea Scrolls.

Among the more vocal critics were Robert Braidwood, a leading American expert in Middle East excavations; Vladimir Milojcic, a professor in Heidelberg, Germany; and Stuart Piggott, who specialized in the Neolithic cultures of the British Isles. All of them pointed out that, in light of traditional archaeological theory, this new dating system made prehistoric dates appear older, sometimes by up to a thousand years. Milojcic went as far as to publish a thorough analysis of the technique in which he questioned its theoretical foundations and criticized its usefulness.

HOW THE METHOD WORKS

Libby had based his method on some physical and biological findings. In the early decades of the twentieth century, physicists had discovered that cosmic radiation, consisting of sub-atomic particles of high energy, constantly bombards the Earth. One of the reactions produced when these particles enter the atmosphere is the creation of radiocarbon, also called carbon-14, a chemical element that behaves in the same way as carbon-12 (the ordinary carbon) but has a different atomic weight. Carbon-14 is rare, with about one atom for every 1 trillion ordinary ones.

Like carbon-12, carbon-14 combines with oxygen to form carbon dioxide, which plants absorb during photosynthesis. Humans and animals eat the plants, or the animals that eat the plants, and all living things end up with the same constant ratio of carbon-14 to carbon-12 in their system.

But when they die, this ratio starts to change because, unlike carbon-12, radiocarbon is unstable, decaying spontaneously, with a half-life of approximately 5,730 years. This means that the ratio of carbon-14 to carbon-12 falls to half in this period. Therefore, by measuring this quotient for, say, linen, charcoal, wood, leather, or bone, we can determine the date when the chemical breakdown began.

Libby started with four assumptions:

- The concentration of radiocarbon in the atmosphere is constant and hasn't changed throughout history.
- The proportion of radiocarbon in all living beings is the same as in the atmosphere and independent of species or location.
- Physical or chemical conditions such as temperature or humidity do not affect the decay of radiocarbon.
- The dated samples are not contaminated, so the ratio of carbon-14 to carbon-12 is not affected by external factors.

In his article against the method, Vladimir Milojcic pointed out that, in general, these assumptions were not justified. Many factors, such as solar activity, volcanic eruptions, or the effects of industrial revolution and nuclear detonations, alter the concentration of radiocarbon in the atmosphere. Also, the contamination of samples is difficult to control. And, finally, measurements are imprecise for very old objects because they contain much less radiocarbon than does the surrounding air. Consequently, special measures must be taken to avoid background tainting.

Several physicists and archaeologists defended the method in the scientific journals *Germania* and *Antiquity.* But a sense of doubt lingered among the experts, and soon Milojcic found a strong argument to support his view. It was based on the discovery of the Tartaria tablets, unearthed in Romania in 1961 (see figure 9.2).

Figure 9.2—The three Tartaria tablets found near Sebes in Transylvania, Romania, in 1961. The circular one is about 6 cm in diameter. The inscriptions on the tablets are among the earliest forms of writing.

These three pieces of baked clay were found in the earth layer of the Vinca culture, which the radiocarbon method had established to be older than the Sumerian one. Since the tablets contained signs resembling the earliest Mesopotamian inscriptions, only one theory was possible: born in Romania, a system of writing spread to the Near East sometime in the fourth millennium BC. But in the 1960s, no archaeologist accepted this idea. It seemed very unlikely that writing had originated in a rural community of Transylvania, and not in the civilizations of the Middle East. This rejection was a blow to the radiocarbon dating method, and its experts started thinking of ways to fix their technique.

Though Milojcic's arguments sounded convincing, in the meantime historians found new interpretations for the Tartaria tablets. One of these theories views the inscriptions in a larger context: this form of writing appeared about 5000 BC along the Danube Valley in southern Hungary, Serbia, Romania, Bulgaria, Macedonia, Kosovo, and northern Greece and flourished for one and a half millennia. Around 3500 BC, numerous tribes invading from the east destroyed these people and their literate culture.

Nevertheless, Milojcic's original criticism of the method was also justified. The factors he had outlined do indeed influence radiocarbon measurements, so the technique did need corrections. Today, radiocarbon results are calibrated with the help of other disciplines, such as dendrochronology, or tree-ring dating.

The controversy, however, has not died. At one end are those scientists involved in developing and applying the radiocarbon dating system and those who benefit from it; at the other stand those who want it discredited. Among the latter are religious groups that oppose the recognition of any date prior to the biblical creation of the world. Their websites emphasize the early blunders of the radiocarbon method. A mistake they are fond of citing concerns a certain archaeological layer found to be a millennium older than a deeper one.

Radiocarbon dating has now come of age and matured into a recognized scientific discipline. Aware of the method's limits, its practitioners never claim to obtain perfect estimates; they always add some margin of error to their results (say, 1,050 ± 60 years old) and attach a probability to their approximations (for example, four times out of five, or an 80 percent probability). Moreover, they never suggest an age after carrying out only a single experiment. Their numbers are released when several independent tests provide consistent readings.

DENDROCHRONOLOGY

Perhaps the most influential dating method after radiocarbon is dendrochronology, introduced by the American astronomer Andrew Ellicott Douglass at the beginning of the twentieth century. While attempting to correlate the size of tree rings with sunspot activity, Douglass created a new science.

A cross-section of a tree trunk shows that the rings vary significantly in thickness (see figure 9.3). In general, each ring corresponds to one year's growth. Size depends on several factors: age (thickness decreases from pith to bark), seasonal temperature and humidity, and what part of the trunk is sampled. Although dry years make narrow rings, excessive rain doesn't make them too large: trees don't drink more than they need to.

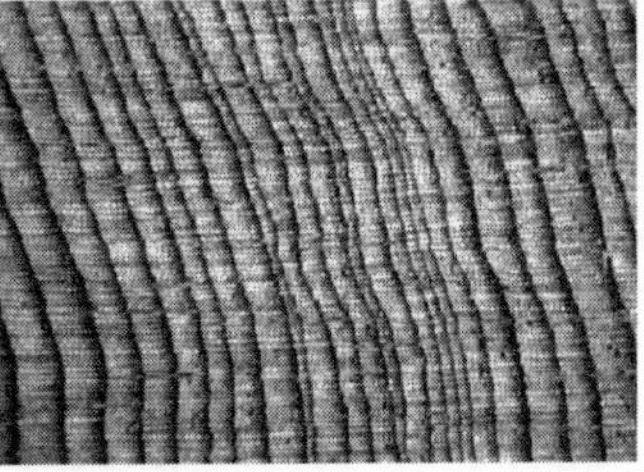

Figure 9.3—The varying thickness of tree rings is the basis of dendrochronology.

By finding overlapping bands between trees of different generations, starting from the living trees, dendrochronologists have catalogued patterns for the last few thousand years. Samples of some construction timber used in Europe, such as the old growth from the Hessian hills in Germany, go back to AD 832. Ring sequences of the giant sequoia reach to the third century BC, and one remnant of the bristlecone pine from the dry White Mountains of California preserves rings created more than nine thousand years ago. Such studies help archaeologists determine the age of communities whose wood has been preserved. One well-known example is the Viking settlement of Haithabu in Schleswig-Holstein, Germany, which has been dated to the tenth century AD.

Tree rings can detect some local weather characteristics, such as very hot and dry periods. Matching the ring pattern with documented reports of past droughts allows a precise dating of those years. For example, the Annals of Lorsch, from the times of Charlemagne, describe an extremely hot summer that killed many people. The width of the rings in the construction oak used in the Saxon village of Hamwih, near Southampton, indicates the year AD 783, which agrees with the historical dating of the chronicle.

Another direct application is to match narrow tree rings with records of unusual volcanic activity. Such a documented event is the 1815 eruption of Tambora in Indonesia, which turned 1816 into "the year without summer." The air was dusty, frosts occurred every month in North America, and many crops failed all over the world. The Sun appeared green in some parts of Asia. Evening skies in England turned red, yellow, pink, and orange. On certain days in New York, the sunspots became visible to the naked eye.

To avoid mistaking thin tree rings corresponding to volcanic activity for those due to drought, researchers also examine the ice-core of Greenland and Antarctica. They can distinguish between layers deposited in different years and identify those containing more volcanic dust. Richard Stothers, a researcher at NASA's Goddard Institute for Space Studies, has linked many such events.

An example of the usefulness of such studies is the year AD 536, for which Stothers found both physical and written evidence of a huge eruption. The corresponding ice layer is very acidic, and tree rings all over the world are extremely thin, but he hasn't yet identified the volcano responsible. Saint Michael the Syrian, the patriarch of Antioch, wrote about that event more than six centuries after it occurred: "The Sun became dark and its darkness lasted for eighteen months. Each day it shone for about four hours, and still this light was only a feeble shadow. Everyone thought that the Sun would never recover its full light. The fruits did not ripen and the wine tasted like sour grapes."

These examples show how dendrochronology helps history find correct dates, but, as expected, the application of the method is limited. Ring patterns depend on both the tree species and the physical location, and scientists are careful about generalizing from local conclusions. And there is a complicating factor: one year doesn't always correspond to one ring.

Some tree species tend to form false rings. For instance, in 1936 and 1937, the Texas yellow pine grew five rings because of early spring frosts. Fortunately, the cell damage is visible under the microscope. More difficult to detect are the rings produced by defoliating insects such as the larch bud moth caterpillar. Others—for example, the spruce budworm—can make rings disappear.

Critics remark that a radiocarbon calibration obtained from a local tree species doesn't apply throughout the world.

Though the criticism is justified, there are ways of verifying the results. One of them is to look at coral deposits. Corals are small animals that live in subtropical waters. When they die, their skeletons form limestone sediments, which grow into reefs, atolls, and islands. Researchers find the age of a deposit by analyzing its layers.

In the late 1980s a group of geologists from Columbia University studied corals raised off the island of Barbados. Using spectroscopy, they calibrated the radiocarbon method back to 30,000 years. Their results confirmed the validity of tree-ring dating for its 9,000-year range. However, this confirmation doesn't solve all the problems of the radiocarbon method, which sometimes disagrees with readings obtained from other techniques.

THERMOLUMINESCENCE

An alternative dating technique is based on the light emitted, in addition to the usual glow, when a crystalline material reaches a temperature of about 500°C. This energy, called thermoluminescence, is stored in crystals after long exposure to nuclear radiation. Pottery contains minerals with high emissions—feldspars, calcite, and quartz. When pottery breaks and shards are buried, the process of building up energy starts again. The quantity of thermoluminescence found in these fragments indicates their age.

Every mineral has a different sensitivity to radiation, and the nature of each type of pottery dictates how mineral measurements and calibrations take place. An objection raised against the method is the experts' assumption that thermoluminescence increases linearly in time. But the charging elements have very long half-lives, of the order of hundreds of millions of years, so the supply of radiation has been practically constant throughout history. Linear growth, therefore, is a reasonable hypothesis.

Specialists know the factors that affect their measurements. In spite of the practical difficulties, they consider their results accurate, with errors no larger than 5 percent to 7 percent, depending on the crystals they deal with. These levels of accuracy seem to be the limits of the method, and there is little hope for better estimates in the future.

Thermoluminescence dating is undertaken when radiocarbon is not applicable or contamination cannot be excluded. Sometimes this technique gives readings very different from several consistent radiocarbon dates. For example, thermoluminescence measurements published in 1972 on the earthenware of Moralba, in Colombia, grouped around AD 1100, a date that disagreed with the radiocarbon results by hundreds of years.

This method is also the better one to use when the researcher suspects a large variation of the carbon-14 to carbon-12 ratio in the atmosphere. Since most cosmic rays don't penetrate into the ground, a doubling of this ratio would increase thermoluminescence by less than 2 percent.

In some instances, radiocarbon and thermoluminescence dating have given similar results, but historians have been reluctant to accept them. In dating the early Jomon Ware found in the Fukuie Cave of Japan, for example, although both methods deemed the samples to be 14,000 years old, archaeologists refused to adopt any date earlier than 3500 BC because it would have contradicted previous findings.

Thermoluminescence can also date some bronze objects. Copper alloys don't store such energy, but their casting cores do. The technique identified the early bronzes of the Chou and Han dynasties in China; Buddhist statues from

Thailand, Cambodia, and Nepal; as well as deity figures from Rome, Greece, Nigeria, and Benin. Until now, historians have largely accepted the dates obtained in this way.

FISSION TRACKING

Archaeology also gets help from particle physics through the fission tracking method. If the atoms of an element such as uranium, which is prone to the spontaneous nuclear disintegration called fission, are trapped inside the crystal structure, the released radiation "scratches" the inside of the rock (see figure 9.4). An electron microscope can detect the marks, whose number provides the age. If the mineral is manufactured glass, the heating used in production erases previous traces, allowing an evaluation of when the sample was made.

Though rare, those minerals are found on archaeological sites, and many objects have been dated by using this method, including ceramic glazes from Japan and glass from Kangavar, Iran. But while more recently manufactured glass is mixed with colorants containing uranium, older samples are purer. For example, some tesserae (mosaic tiles) found in Pompeii show only minor traces of radioactivity.

Sometimes, fission tracking disagrees with radiocarbon readings. For instance, the former method found that a vase

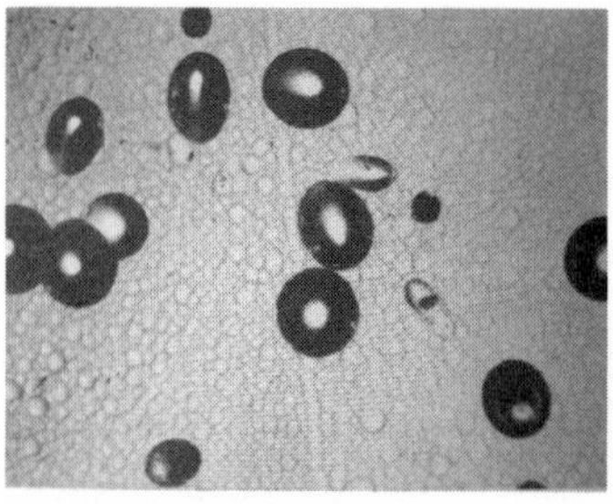

Figure 9.4—Spontaneous fission tracks left by uranium-238 in glass. This image is magnified; in reality, the marks are about 0.002 mm long.

unearthed in Tosamporo, Japan, was between 4,680 and 5,480 years old, while the latter placed it at around 3,950 years old. Also, a glass sample from Landberg Fort, on the northern coast of Scotland, was deemed to be only three centuries old and not from the Iron Age, as previously thought.

But often the two methods agree. An obsidian knife blade, found in a cave along the Rift Valley in Kenya, has a fission track date within the radiocarbon range for objects found at the neighbouring Njoro River Rock Shelter. Also, there is agreement between the fission tracking and the radiocarbon ages obtained at the archaeological site of Olduvai Gorge in East Africa.

ARCHEOMAGNETIC DATING

Another method used for the time range of history is archeomagnetic dating. Its goal is to establish the age of objects by comparing their magnetic information with changes in the Earth's magnetic field.

A magnetic force, whose direction and intensity change in time, acts at every point on the globe. The variation is due to the Earth's core, which is rich in iron and nickel, and the complicated motion of the Earth's axis, a component of which is the precession.

Certain types of minerals retain the magnetic information that was present at the time they came into existence. Only temperatures exceeding the so-called Curie point, which is rock dependent, can erase this record. (Magnetite, for example, has a Curie temperature of 580°C, and hematite, 680°C. Note that wood fire alone doesn't exceed 500°C.) While cooling again, the minerals register the magnetic imprint of the place at the given time.

Figure 9.5—A map of the intensity of the Earth's total magnetic field in 1980.

The first task of the experts was to draw detailed magnetic maps, which give the direction and the intensity of the field at every point on the Earth's surface and at different periods (see figure 9.5). Work started in 1925 with mapping the area around Mount Etna in Sicily from various layers of lava, previously dated from AD 1284 to 1911. The early rock studies began in Germany and the United States in 1938, and the method advanced rapidly after the invention of the radiocarbon technique.

An essential practical requirement of archeomagnetic dating is not to change the initial position of the measured object. Among the materials suitable for testing are fixtures such as kilns, burned wall surfaces, central-pit house hearths, and fire-hardened floors. One problem the expert must always solve before proceeding with collecting the data is whether the sample mineral has reached the Curie point. If it hasn't, the material's original magnetic imprint will interfere with the new one, leading to incorrect results.

As with other techniques, archeomagnetic dates haven't always agreed with those of radiocarbon. That was the case, for example, with the measurements Daniel Wolfman of the Museum of New Mexico took in the 1970s and 1980s. Many of the tests he conducted in western Honduras, El Salvador, Belize, and Mexico led to results different from the established ones, even if only by a few hundred years. Nevertheless, he thinks that this part of the world now has a relatively good chronology for the period AD 1–1200.

Radiocarbon and traditional history influence archeomagnetic dating because charts cannot be drawn without relying on some established chronology. Still, the results this method obtains at new sites do not always match those of radiocarbon.

FOMENKO'S EVALUATION

Fomenko's criticism of the scientific dating methods mentions disagreements similar to the ones discussed above. Whereas his earlier work attacked only the radiocarbon technique, dismissing it in short order, his more recent book *History: Fiction or Science?* takes the problem more seriously. In almost twenty pages, Fomenko explains why a correct chronology could not be constructed on the basis of the scientific methods. But most of his arguments are founded on the early failures of Libby's radiocarbon technique.

He cited a 1970 article, published in *Nature*, about the mortar used to build an English castle known to be 738 years old, which the radiocarbon method misdated by 6.5 millennia. He also quoted a 1971 issue of the *Antarctic Journal of the United States* concerning the age of some recently shot seals, placed by the method in the eighth century AD. Another argument was from a 1979 article published by the Manchester Museum, whose curators dated a mummy to 1000 BC and its clothes to 380 AD.

It is difficult now to convince a scientific audience with such examples. The early radiocarbon blunders are well known, and experts no longer make generalizations after one measurement. Some researchers go as far as to ignore works published on the subject before 1980.

Therefore, it is surprising to read in Fomenko's latest book a story about a 1989 experiment, organized by Britain's Science and Engineering Research Council, in which thirty-eight radiocarbon laboratories from all over the world were involved in a precision testing of the method. Fomenko described the outcome as follows: "[The laboratories] received specimens of wood, turf, and carbonate salts whose age[s] had only been known to the organizers . . . but not to [the] actual analysts. Only seven laboratories (of thirty-eight!—A.F.) reported satisfactory results; *others proved wrong by factors of two times, three times and higher.*" [Italics in original]

This passage leads the reader to think that if the correct age of a sample was, say, 1,000 years, some measurements provided readings of 2,000 years, 3,000 years, or even more. Could that happen in 1989, after four decades of progress, billions of dollars spent, and a Nobel Prize awarded?

A half-page article entitled "Unexpected Errors Affect Dating Techniques," signed by Andy Coghlan, had indeed appeared in the September 30, 1989, issue of the *New Scientist.* Coghlan had interviewed Murdoch Baxter, the director of a research centre near Glasgow and the organizer of the experiment. Baxter said that only seven out of thirty-eight laboratories produced satisfactory results, while the others were "two or three times less accurate than implied by the range of error they stated."

But this is not what Fomenko wrote. According to Baxter, if the correct age was, say, 1,000 years, some laboratories produced a result of 1,080 ± 40, whose estimated error should have been ±80 (twice bigger) to include the correct date. Others had such readings as 940 ± 20, whose estimated error must have been ±60 (three times bigger) to capture the truth. The meaning was clear from Coghlan's first sentence, "The margin of error . . . may be two to three times bigger," which doesn't mean that the results proved wrong by factors of two, three, or more, as Fomenko claimed.

The article states that while a few laboratories dated all their samples almost to the year, others were off by up to a quarter of a century. That means the mistakes had no connection with the technique, and even the largest errors were nothing like those that occurred during the method's pioneering days. But, unfortunately, Fomenko omitted these details.

The British experiment had an impact on radiocarbon dating centres all over the world. The International Atomic Energy Agency decided to improve the standards of the reference samples used for testing the machines and revised the technology for the chemical pre-treatment of the samples. Still, the critics continued to oppose the method.

Other arguments Fomenko brought against the scientific dating techniques came from several books by the German authors Christian Blöss, Heribert Illig, and Hans-Ulrich Niemitz with such titles as *The Circular Arguments of Dendrochronology*, *C-14 Crash*, and *The Self-Deceit of the C-14 Method and Dendrochronology.* But academics dismiss these independent researchers of history as being opponents of Scaliger and Petavius in spite of lacking scientific and archaeological expertise. So the German authors may be biased as well.

Fomenko included in his book an article entitled "A Critical Analysis on the Hypotheses on which the Radiocarbon Method is Based," written by the mathematician A.S. Mishchenko. Though the piece is thoughtful and well written, it stresses only the known shortcomings of the technique and does not touch on its merits.

After reading Fomenko, the uninitiated reader is left with the impression that a large number of scientists are involved in a hopeless enterprise. On the one hand, he maintains, the results they offer can be very bad; on the other, traditional chronology has tainted the calibration of the methods, leading to erroneous results.

This picture is deceptive. The scientific readings do include errors—but they are not as large as Fomenko thinks, and the readings' precision is improving. Also, the calibration has nothing to do with Scaliger's chronology. Though tradition might influence the assessments of historians and archaeologists, this fact doesn't change the measurements obtained through scientific dating methods. Another important aspect Fomenko overlooks is the statistical nature of the results, which are viewed not individually but within a group. If the numbers obtained are inconsistent, the experiment is dismissed. This measure strengthens the method.

OTHER TECHNIQUES

Around 1980 yet one more technology was invented to assist in dating ancient materials. The accelerator mass spectroscopy technique has many advantages over the classical way of measuring the ratio of carbon-14 to carbon-12. This method needs only tiny samples of between 1 and 3 milligrams, rather than the 30 grams Libby's technology required. Parchment or papyrus manuscripts that had never been checked before could now be tested.

In 1995 a team of experts from the University of Arizona in Tucson analyzed the Dead Sea Scrolls of the biblical book of Habakkuk. They dated several fingernail-size fragments of parchment, papyrus, and linen to between 150 and 5 BC, with a probability of 95 percent—in good agreement with what a Swiss laboratory had obtained in 1992 for similar samples but with a confidence of only 68 percent.

Douglas Donahue, a physics professor and the director of the accelerator mass spectrometer that did the Arizona measurements, explained that the university team's procedure was faster and more reliable than others. As a result, between 1980, when the laboratory started functioning, and 1995, when it tested the scrolls, his team had dated more than twelve thousand objects. Among them was the Shroud of Turin, which proved to be of medieval origin.

The Dead Sea Scrolls were also tested with other methods, such as paleography, which analyzes handwriting. The dates resulting from this check proved similar to the ones provided by Arizona. But some critics have attacked this kind of research. One of them is Robert Eisenman, a professor at the California State University at Long Beach, who explained in 1983 that paleography can at best order events but has no means to provide absolute figures. In his view, the Dead Sea Scrolls are the work of early Christians.

Another technique processes the collagen fibres of the scrolls made of animal skin and compares them with fibres of known ages. Though this method cannot provide absolute figures either, it showed that the original Dead Sea Scrolls are older than the parchments found near Qumran in a Wady Murabba'at cave, and which bear written dates from AD 132 to 135. All these results add to the credibility of the radiocarbon method and the other scientific dating techniques.

Clearly, Fomenko's criticism shows only one side of the coin. Instead of performing a careful analysis of the available dating techniques, he builds an intelligent case to discredit them. A careful look at his references, however, reveals that he relies on sources that agree with him but overlooks those that don't.

In every branch of knowledge, the methodologies and the views about them evolve. During the late 1960s it was said that if a carbon-14 date supports archaeological claims, then it goes into the textbook; if it doesn't contradict them, it becomes a footnote; and if it opposes the historians' belief, it should be ignored. Today the results are more trustworthy, and radiocarbon dating cannot be discarded as nonsense.

SUMMING UP

As most specialists admit, the scientific dating methods are not independent of each other. The connections between them show that none has the capability to offer precise absolute dates for ancient and medieval history. Their results are approximate, and they rely on external findings. In other words, they rest on each other.

In principle, this approach could lead to a vicious circle, and authors such as Heribert Illig have exploited this possibility. Also, most experts know the limits of their conclusions, and disagreements among them are common. But, as in any active field of science, the scientific dating methods make progress.

Dendrochronologists continue to study more and more old trees to help improve their calibrations of the radiocarbon technique. Often they take into consideration thousands of samples from the same species. Thus, the criticism that the bristlecone pine and the sequoia of North America cannot reveal much about radiocarbon levels in Europe loses ground, and the scientific methods gain more and more acceptance.

Though not absolutely accurate, the dating methods can provide an approximate answer to the chronology questions investigated here. In a way they act like opinion polls before elections: They are reliable within a certain margin of error. They are not very useful when the race is tight, but they do not predict the opposite outcome when the balance of forces is clear.

Are these methods demolishing Fomenko's doctrine? Not obviously so. No scholarly work addresses the relationship between his theory and the results of scientific dating techniques. An analysis of this sort would be difficult to perform. The radiocarbon date of some archaeological object doesn't necessarily contradict a king's years of reign: the object may not have belonged to that king. History often associates disparate events through traditional chronology, so a study of Fomenko's proposals must rely only on findings that are independent of any such connections.

A good way to approach this problem would be to use the radiocarbon method for dating old manuscripts whose provenance is certain, as was done with the Dead Sea Scrolls. That would help establish the relationship between scientific dating results and Fomenko's astronomical and calendrical findings. Do some radiocarbon measurements contradict his conclusions? Do any of them agree with his results? It is time to open up the field.

CHAPTER 10

Finding a Consensus

I like the dreams of the future better than the history of the past.

THOMAS JEFFERSON

Anatoli Fomenko did not convince me about the correctness of his shift theory. He had, however, raised some good arguments against tradition. But they were based on interpretations different from those of conventional chronology. I had to decide what drives these choices and whom I could trust.

While reading the works of renowned historians and talking to experts in antiquity, I noticed that they had few or no worries about the correctness of their dating system. Most of them took it for granted. My colleague Gregory Rowe from the Department of Greek and Roman Studies at the University of Victoria paraphrased the Cambridge historian Keith Hopkins in saying that the evidence of ancient history is like the rods holding an Indian tent: none of them stands by itself, but they support each other.

Though I liked this comparison, it doesn't explain Fomenko's astronomical objections, including those related to the Peloponnesian War. As explained in chapter 4 of this book, the reliance on Thucydides' eclipses makes this war a crucial chronological landmark. So I kept thinking about ways to find out which of the three possible dates (431 BC, AD 1039, or AD 1133) marks the beginning of the war. Answering this question would have been a good start towards refuting or accepting the traditional dating system.

One day an idea flashed through my mind, and I asked Greg Rowe for help. Fomenko had fixed the beginning of the war to AD 1039 and the First Council of Nicaea to AD 876. The traditional dates are 431 BC and AD 325, respectively. Since Fomenko had changed the order of events, I wanted to know if Greg Rowe could give me a good and simple reason why the reversal made no sense. This is what he answered:

> I'm sorry I can't give you a good and simple reason. What you'd like, I think, is a text in which someone known to have been present at the Council refers unambiguously to the Peloponnesian War as a prior event. There may be such a text: we have voluminous writings of numerous bishops who were present; or there may not: the Peloponnesian War had occurred some seven centuries earlier, the war itself was less significant than the fact that Thucydides wrote a history of it, and the Nicene fathers were preoccupied with the question whether God and Jesus were of the same, or only of similar, substance. In any case, I don't know of such a text. Nonetheless, a clear sequence of events brings us from the Peloponnesian War to the Council of Nicaea. This sequence involves, for

> example, Alexander, Rome, Jesus and Constantine. It is less well known in some parts and better known in others. But its existence is beyond any doubt.

Rowe's hint gave me hope. Though the directions he proposed would have provided me solely with a relative chronology, I had only three dates to choose from. So the mention of the Peloponnesian War at the Council of Nicaea or the existence of Rowe's sequence would have tilted the balance of probabilities towards tradition. But all my efforts to fill in the details from primary sources failed. I found no convincing link between the two events. That may be so only because of my inability to navigate inside the labyrinth of documents; still, whatever the reason, I lacked a proof.

BACK TO THUCYDIDES

In the fall of 2003 I learned that the State and University Library in Hamburg, Germany, housed some papyrus scraps, dated to the third century AD, which had come from a copy of Thucydides' book. Further inquiries led me to an expert, the papyrologist Dieter Hagedorn of Heidelberg, whom I asked for details. He told me that the fragments had been paleographically dated (see chapter 9), and he also helped me contact the Hamburg library. As far as he knew, none of the other papyrus copies of the original—located in museums or archives in various parts of the world and dated from the first to the seventh century AD—had ever been radiocarbon tested.

The next day I called Douglas Donahue, the Arizona physicist who had carbon dated the Shroud of Turin and the Dead Sea Scrolls (see chapter 9). Though retired, he was still active in his laboratory and agreed to test the fragments for me. The only thing I had to do was mail a sample to him.

I needed from Hamburg a 3-milligram piece of papyrus, not larger than a fingernail, so I wrote the curators a letter, requesting a sample in exchange for a radiocarbon date. I made it clear that I would take care of all the expenses, which Donahue had estimated at $400. The answer came a week later from Eva Horvath, the head of the library's manuscript section. Her message was polite but firm: no way.

I insisted. Would she help me approach her boss to consider my request? Horvath expressed regrets. She could do nothing for me. The papyrus was too fragile; the curators kept it away from light and humidity; and the head of the restoration department would never break the rules.

Was it worth trying further? I decided it was not. What could a single test prove? Carbon dates are taken seriously only after several independent checks. And one manuscript alone was not enough either. What about the other ninety or so copies? To get a meaningful answer, I would have to check them all. And even if I had obtained the samples, I couldn't afford the cost of the procedure.

But in making the attempt, I glimpsed a possible solution: perhaps a research institution or a private foundation would help me move ahead. What questions should this project answer? I saw three potential outcomes:

- Most radiocarbon readings point at conventional dates. This result would confirm the findings of history and consolidate the traditional system.
- Most readings are not much earlier than AD 1000. This finding would give a boost to revisionist claims, questioning not only the chronology of ancient Greece but also the framework of papyrus dating.

- The readings are spread along more than a millennium. If so, then a statistical analysis would (at best) provide some answers or (at worst) be inconclusive.

I haven't changed my mind since formulating this solution: the plan is worth pursuing. It might explain some of the problems raised in this book either by proving them illusory or by showing that historians must review the distant past. But the experiment could also fail. What then? Are there other ways to learn the truth?

HISTORY AND INTERPRETATION

Between 1934 and 1961 the British historian and philosopher Arnold Toynbee published his twelve-volume treatise *A Study of History*, in which he compared twenty-six civilizations, analyzing their creation, growth, and fall. The book had a strong influence on the modern attitude towards history, religion, and international affairs and left its imprint on the way historians perceive their field.

Toynbee began his masterpiece with a chapter on the relativity of historical thought. To illustrate his idea, he invoked a comment by the Greek philosopher Xenophanes (*c.* 570–475 BC):

> Ethiopians make their gods snub-nosed and black, and Thracians make theirs blue-eyed and red-haired. If only oxen and horses had hands and wanted to draw with their hands or to make the works of art that men make, then horses would draw the figures of their Gods like horses, and oxen like oxen, and would make their bodies on the model of their own.

Toynbee's point was that different cultures, like different individuals, perceive reality with different eyes. Historical thought is relative and follows the tendencies of its time. By showing that the ancients understood this principle, Toynbee stressed both the ubiquity and the natural character of this truth.

This recognition leads to the question of how different two interpretations of the same source can be. Obviously, the source plays a crucial role. The article written by a journalist on the recent invasion of Iraq may leave less room for doubt than a fragment from Herodotus about a battle he hadn't seen. Still, each piece of writing can be viewed in the context of corroborating evidence, and various interpreters may draw different conclusions from what they read. But how much can their opinions vary?

Like any other field of inquiry, history has developed its own canons, standards, and ways of thinking. These benchmarks may change over time more than they do in the exact sciences, but modern students of history are not going to differ too much from their colleagues, even if they often disagree. They have been educated to think similarly. They have learned to accept certain facts and reject others.

No historian of antiquity, for example, would contest the 50 BC estimate provided by the French Egyptologist Sylvie Cauville for the round Denderah stone, give or take a few years (see chapter 6). Cauville had initially placed its carving between 51 and 43 BC, using historical arguments, and later arrived at the date of 50 BC by analyzing the positions of Mercury and Mars relative to constellations, ignoring the other planets of the horoscope.

On the one hand, Cauville's assessment agrees with the overall scheme of Egyptology; on the other, the fine-tuning of the date is based on false astronomical reasoning. But historians would prefer to discard the astronomical evidence (perhaps going so far as to say that the Denderah stone is a mere work of art) than to contradict the historical framework pointing at the mid-first century BC. Otherwise, the entire system of standards on which Egyptology is based would collapse.

In other words, historians take the existing chronological system for granted and discard any information that disagrees with it. Except for a few "problem" events, such as the founding of Rome or the ascension of the Egyptian king Menes, for which they admit to having difficulties with fixing the date, contemporary historians do not question traditional chronology and discard any information that contradicts it. Then, of course, the pieces come together, and the "Indian tent" in Greg Rowe's analogy stands.

In chapter 5, I presented an extreme case that illustrates this attitude. In his study of the Moon's acceleration, Robert Newton accepted only a handful of the 370 eclipses he surveyed, complaining that "we have found too many instances of an eclipse that could not possibly have been total but that was so recorded, sometimes in a quite picturesque manner." Why would so many observers over the centuries make the same mistake? Isn't the elimination of these testimonies (for the sake of accommodating traditional chronology) as narrow-minded as Fomenko's ignorance of some dated Russian texts (see chapter 8)?

Fomenko adheres to a different set of values from that of historians. To him, celestial mechanics and astronomy have priority, and historical arguments contradicting scientific results should be dismissed. As a mathematician, he feels no attachment to the canons of history and is insensitive about destroying an edifice that many historians have built over several generations. But what he fails to accept is that, by modelling history along new lines, he creates bigger problems than the ones he can solve.

Christopher Mackay pointed to this difficulty (see the introduction): by switching the dates of the Peloponnesian War and the Battle of Pydna, Fomenko unknowingly implied that Alexander the Great didn't exist. So what is easier to accept: that the Macedonian king was an invention, or that Thucydides erred about stars being seen during the first eclipse of the war (see chapter 4)?

No doubt, every historian would choose the latter option, while Fomenko would say the question is unfair. He never claimed that Alexander the Great had been invented; he merely doubted conventional chronology. If that is what tradition leads to, it does so at its peril. History must explain both Alexander and Thucydides in the context of the documented evidence. But if this evidence points at the reversal of the events, historians must reconsider the data and revise their interpretation.

In the end, for historians and reformers alike, the assumed chronology is the guiding force behind historical interpretation, the reference frame driving the choices. From Scaliger and Petavius to Newton and Fomenko, every inquirer has fixed a few dates and tried to connect them to the events mentioned in the chronicles. To make sense of those stories, the researchers have dismissed some facts and accepted others.

The choices one makes depend on one's values and beliefs. The first part of this chapter, for instance, illustrated my trust in the power of the radiocarbon method to draw statistical conclusions and provide approximate dates. My choice was not random, and chapter 9 explains why. Similarly, historians on the one hand and reformers on the other have their reasons for taking a certain path.

These examples indicate how important the historian's priorities are. What comes first: the astronomical evidence, the word of the chroniclers, the legends, the sacred texts, the scientific dating methods, or some combination of them all? This question cannot be answered without considering another question: What qualities must a historical source possess in order to be accepted as legitimate?

THE BIBLE: HISTORY OR MYTH?

In 1955 German journalist Werner Keller published a book entitled *The Bible as History.* It became an instant success, was translated into twenty-four languages, and sold more than 10 million copies all over the world. Keller argued for a historically accurate Old and New Testament, basing his account on archaeological evidence. Guided by his trust in the veracity of the Scriptures, he ended his introduction with the statement: "The Bible is right after all!"

But he hadn't always believed that. Before learning about the archaeological discoveries in the Middle East, Keller viewed the biblical reports as fiction. What changed his mind was his detailed review of the many relevant history articles published in specialized journals. Their conclusions fascinated him so much that he decided to blend them. Thus, one of the best-selling books of its time was born.

Keller started with evidence endorsing the patriarchs, continued with accounts of Egypt from Joseph to Moses, and followed the journey to the Promised Land. He analyzed every book of the Bible using the latest discoveries, most of which endorsed his point of view. When some findings were inconclusive, as happened with the Shroud of Turin, Keller honestly admitted the fact. Nevertheless, in a later edition of his book, he expressed the hope that a radiocarbon test would clarify the issue.

But not everybody agreed with the idea of a factually true Bible. Northrop Frye, who died in 1991 after a brilliant career as a literary theorist and professor at the University of Toronto, thought the Bible was historically accurate only in places and only by chance. Tom Harpur, a journalist, theologian, and teacher of Greek and New Testament at the same university, shares this view today. Moreover, Harpur believes that Jesus didn't exist (see chapter 4).

In the year 2000 the British historian, writer, and documentary filmmaker Ian Wilson published *The Bible Is History*, which he intended as a sequel to Keller's book. As the title shows, Wilson was inclined to agree with Keller, but he approached the problem objectively, emphasizing its controversial aspect. His account provides an image of the antagonistic views held in the field of biblical scholarship.

A recent contribution to the debate between those who accept the truth of the Christian Scriptures and those who don't is a book from 2001, *The Bible Unearthed: Archaeology's New Vision of Ancient Israel and the Origin of Its Sacred Text*, by Israel Finkelstein, chair of the Archaeology Department at Tel Aviv University, and Neil Asher Silberman, a historian and journalist. The two experts think the Bible is pure mythology.

In their opinion, the Israelites had not come from Egypt, did not wander in the desert, and never conquered the land of Canaan. The kingdom of David and Solomon, described in the Bible as a political and military power, was a small tribal community, and Jerusalem had not been the capital of an empire but an ordinary town.

The authors claimed no originality; biblical scholars had accepted many of those facts years before. Among them are the anachronisms and contradictions they had discovered in the tales of the patriarchs Abraham, Isaac, and Jacob, which, in the scholars' opinion, prove these accounts false. Historians had to conclude the same about the Exodus after extensive archaeological excavations found no evidence of large encampments in the Sinai Peninsula, where Moses allegedly brought the Israelites after crossing the Red Sea.

Conservative Israeli historians and politicians reacted angrily to this book. They accused the authors of "historical minimalism" and of questioning the legitimacy of the Israeli state. But Finkelstein and Silberman stood firm. The Bible is not history, they maintained, and they refused to apologize for telling the truth.

This episode is a clear illustration that the experts sometimes disagree on what is an acceptable historical source. Some of them trust the reality of biblical accounts, whereas others don't. A good deal of currently accepted ancient history, however, is based on the Bible, so those who doubt the veracity of the Scriptures also question the correctness of traditional chronology.

The lack of consensus in accepting a certain source as historical evidence may lead to radically different chronologies. But unanimity alone is not enough either, because the source agreed upon may be capable of various interpretations. That leads to the question: What mechanism drives the choice to accept one fact as historical and another as fiction?

CHOICE AND CREDIBILITY

Choice is often based on credibility. Credible people are not necessarily those who seek the truth, but those who persuade their peers. Others may be correct but unconvincing. When

Columbus, for instance, claimed that the Earth was round and that by sailing west he could reach India, very few believed him. Though he was not credible, Columbus was right.

When choosing to accept a fact as historical or to dismiss another, historians must sound credible to other historians, whose opinion matters to them. They don't have to convince writers, chemists, or mathematicians. The statement that Thucydides dramatized when mentioning stars during the Peloponnesian War's first eclipse, for example, may sound very reasonable to historians, though Fomenko rejected it.

Gaining credibility is often more important than finding a solution. Remarkable in this sense was the attitude of the famous historian Theodor Mommsen, who, in his book on Rome's chronology (see chapter 2), encountered the following problem. Around the middle of the second century AD, Emperor Antoninus Pius celebrated Rome's nine hundredth anniversary, in agreement with the traditional date of 753 BC. Two centuries later, a similar festival took place to honour the city's millennium. But the numbers don't add up; the millennium should have been marked a century earlier.

It's not very likely that, in two hundred years, the memory of the previous anniversary was lost. But sometimes mistakes happen, as they did with the celebrations to welcome this millennium, which took place in the year 2000 instead of 2001. Given that there is no year 0 and the Christian era started in AD 1, the second century began in 101, the third in 201, and so on. Though the arrival of the twentieth century had been correctly marked in 1901, nobody seemed to remember that a few years ago.

Does this anomaly mean that, in AD 3000, historians will think that a year is missing from the twentieth century?

Unlikely. They may have enough evidence to draw the right conclusion. In his case, Mommsen "solved" the problem by noting that the beginnings of Rome are nebulous, and that the two celebrations probably relied on different sources.

Mommsen's explanation sounds reasonable, and though it is given at the expense of losing the legitimacy of the year 753 BC as Rome's birthdate, it is a natural choice. But Anatoli Fomenko sees this accommodation as an admission of weakness, citing Mommsen as endorsing a 500-year divergence between the various estimates for the founding of Rome. No doubt Fomenko is arithmetically right, but history is not mathematics.

Sometimes Fomenko leaves the impression that he understands the nature of ancient and medieval history, whereas at other times he conveys the opposite. His approach varies with his target readership. In a treatise aimed at experts, for instance, Fomenko is cautious about the role mathematics can play in this field:

> The work discussed here cannot claim to be the basis for any final conclusion, the more so as the most complicated, multifarious and often subjectively interpreted historical data are analysed by strictly mathematical methods. To process the material will certainly require a large variety of methods: purely historical, archaeological, philological, physical and chemical, and . . . mathematical, which . . . will permit us to look at the problems of chronology from a new angle.

But in books addressed to an educated general public, he takes the authoritarian position that Zalyzniak denounced (see chapter 8):

> We believe that the unprejudiced reader is already convinced that we are moved by the firm logic of scientific research . . . We have to move further along this path if we want to stay on the ground of common sense and strict scientific principles . . . [Our conclusions] follow directly and unambiguously from medieval Russian documents . . . The identification of Russia in a certain period with Ireland follows unequivocally from old English chronicles.

Fomenko's credibility is a delicate issue, and opinions vary. Usually each critic refers to a different side of his personality: mathematician, historical reformer, or visual artist. But even his chronology work has led to contradictory views. Zalyzniak and Novikov, for instance, criticize him harshly (see chapter 8); others are at the opposite pole.

Among adherents is A.N. Shyraev, for many years the head of the Probability Department at the Moscow State University. In 1990 he wrote that Fomenko's "ideas are perfectly rational from the point of view of contemporary mathematical statistics and fit into the cognitive paradigm of experts." In the same spirit, the logician and sociologist Alexander Zinoviev praised Fomenko's work for its "exceptional scientific scrupulousness."

But with more critics than supporters, Fomenko is not in a comfortable position. Like any scholar swimming against the academic current, he pays dearly for his mistakes. It would be wrong, though, to treat his work superficially and place him at one extreme or the other. His ideas are entitled to a careful analysis.

Fomenko's chronology results seem to fall into three categories: good, mediocre, and blunders. The first have enough credibility to merit serious attention; the second are set on a shaky foundation, but their overall framework is worth investigating; the third have damaged his academic reputation and continue to harm him.

Among the good results obtained alone or with his team are those concerning eclipses, the *Almagest*, and the dating of the First Nicaean Council. In all these cases, Fomenko followed the path of previous chronologists and, by making reasonable but not identical choices, reached very different conclusions from those of Scaliger and Petavius.

Do these achievements require a rewriting of history? The astronomical arguments in favour of tradition are strong. The works on this subject by experts such as Theodor von Oppolzer, Friedrich Ginzel, Otto Neugebauer, Robert R. Newton, Justin Schove, and many others support the established chronology. It is true that these researchers made choices, accepted some celestial records and ignored many others, but they did their homework. If Fomenko wants to confront them, he needs to accumulate at least as much evidence in his favour. However, this road is not a dead end for him.

Among Fomenko's mediocre contributions are those on statistical methods for texts and maps, overlapping dynasties, and the dating of Egyptian horoscopes. Overall, they lack rigour, and many of their conclusions contradict the historical evidence. But these techniques are ingenious and deserve to be studied. Some could be developed into useful tools—for example, those for texts and maps. Others might prove hopeless, as is likely to happen to the dynastic method.

An obvious blunder is Fomenko's etymological analysis, which ignores the basic rules of an established discipline. As Zalyzniak argued, Fomenko's approach violates the fundamental principles of linguistics. Fomenko has made childish mistakes in this area, ones he could easily have avoided by consulting the literature.

Another error Zalyzniak denounced was Fomenko's belief in a systematic forgery of ancient and medieval documents led by a conspiracy plot that continued over the centuries. The reasons Fomenko gave for this theory are summarized in a quotation from George Orwell's *Nineteen Eighty-four*: "Who controls the past controls the future. Who controls the present controls the past." There is no doubt that some documents are forged for political purposes. But generalizing from isolated acts to a systematic long-term enterprise is a huge leap, and the difficulty, energy, and motivation involved in doing this might overwhelm the richest imagination.

All these considerations converge to a natural question: How is it possible that a capable man appreciated in his field, who leads a research institute for mathematics and is a member of a respected academy, can, simultaneously, bring intriguing contributions to crucial problems and make naïve errors of judgment? The answer is less complicated than it seems.

MATHEMATICS AND SOCIETY

Almost every field of human endeavour has a public image, and mathematics is not exempt. Neither are mathematicians. They are perceived differently at particular times and in various parts of the world. In North America today, this image is shaped by some stage productions and Hollywood movies that portray them as inflexible, weird geniuses who often live on the verge of mental breakdown.

Nothing could be further from the truth. Though such individuals exist, I would be surprised to see a study proving them to be more numerous than in other professions. Mathematicians are like anyone else; it's just that they enjoy mathematics.

In other parts of the world, the image is different. China, for instance, treats mathematicians with pious respect. At the International Congress of Mathematicians held in Beijing in 2002, wearing the conference badge earned the participant not only free public transportation but also admiring looks from the passers-by. President Jiang Zemin opened the meeting in the Great Hall of the People and offered a dinner to the six thousand participants in the banquet hall that had hosted Richard Nixon three decades before. In contrast, the highest political presence at the congress of 1986, held in Berkeley, California, was the city's mayor.

These extremely different attitudes may be explained by the mystery surrounding mathematics, at whose door many knock but few may enter. Mathematicians have been viewed sometimes as wizards and at other times as witches. This is not just a metaphor; Saint Augustine of Hippo (*c.* 345–430), one of the four fathers of the Latin church, wrote in *De Genesi ad Litteram* (The Literal Meaning of Genesis): "The good Christian should beware of mathematicians and of all those who make empty prophecies; the danger exists that they have made a pact with the devil to darken our spirit and confine us in Hell."

Psychologists know that the national perception of each profession influences its practitioners, from helping them grow confident if the image is good to inducing an inferiority complex when the public mood changes. To understand Fomenko, we have to know how mathematics was perceived in Russia during the last third of the twentieth century, when he developed most of his ideas.

In many ways the attitude resembled that in China today. Rigid and autocratic, the Soviet regime promoted the fields endorsing the Leninist world view. It marginalized religion, but it funded the sciences heavily. No wonder the Russians were the first to launch a satellite, *Sputnik I*, in 1957 and to put a man, Yury Gagarin, in space in 1961.

These accomplishments would not have been possible without mathematics, which was called the "Queen of the Sciences." But while some Soyuz rockets exploded and some physical theories failed, mathematics flourished. Children learned in school that Pythagoras' theorem was true two millennia ago, is true now, and will stay true forever. This feeling of stability gave mathematics the key to the universe, whose secrets could be reduced (at least in principle) to equations and formulas. Mathematics ruled the world.

Mathematicians were loved by some and envied by many. They belonged to the most intelligent breed on the planet; they were esteemed above scientists, and well above literary critics, painters, historians, and writers. Physicists were failed mathematicians, chemists were failed physicists, and so on. But mathematicians didn't care much about this hierarchy: they were at the top, so why waste time looking down on mortals?

This nonsense loomed large in the Eastern Bloc. Because everyone, except for a few Communist Party officials, was equally poor, people sought idealistic means to feel superior, and mathematicians happened to live in a lucky environment. But many of them grew haughty. The more experienced club members held the view that those capable of developing a mathematical way of thinking could switch direction and climb to fame in any field they wished to enter. They could also become physicists, linguists, historians, or writers, and the effort to reach those lesser peaks would be minimal.

Of course, not all mathematicians believed that line. Some looked around them, tasted defeat in other directions of human endeavour, and understood that they couldn't succeed everywhere. But most of them were busy developing their domain and competing with minds comparable to theirs. Why would they care about the natural order of things?

Now, when the realities of a free economy have ravaged the system of values that had been established in the Soviet society, the attitude is changing. Money and corporations are starting to play a leading role in eastern Europe. But the old concepts are not dead and will probably breathe for another decade or two, until the younger generation takes the lead. Sooner or later, mathematicians will have to adopt the humbler attitude of their Western colleagues, who, like everyone else, fight for the survival of their profession.

The way mathematicians were viewed in the Eastern Bloc, and especially the way in which they regarded themselves during the time Fomenko developed his ideas, can explain many things. So I am not surprised about his linguistic blunders and suspect that he fell into this trap because of a philosophy that puts mathematics above other disciplines.

Zalyzniak's reaction to Fomenko's proposals also fits this view. Though well argued, his account showed both the frustration and the satisfaction of a man who could, if not dispel, at least deal a heavy blow to a vexing myth. His reasons for adopting an insulting tone may be deeper, but this possibility doesn't negate my view.

Fomenko's blunders may also have been triggered by his success with the Russian public. The documentary movies he made and the media interviews he gave may have contributed to his growing confidence and increasing lack of rigour. His work on Egyptian horoscopes, for instance, became less scrupulous over time and suffered from weaker interpretations in later years (see chapter 6).

More recently, Fomenko made the mistake of putting absolute trust in his chronology. A too strong belief in a system can weaken the critical spirit and make the proponent accept statements without giving them enough thought.

But Fomenko is not the only one who has blundered. Historians have made mistakes too, and the previous chapters have identified some of them. In general, their errors are less apparent because they reinforce the status quo. Sometimes, however, they are more than obvious. A typical example has been provided by another historical reformer, Gunnar Heinsohn, a professor at the University of Bremen and the director of the Raphael Lemkin Institute for Xenophobia and Genocide Research.

Heinsohn referred to the archaeological findings at Hazor, an ancient Canaanite city in northern Israel. In 1996, excavations made in the upper strata of the site uncovered four cuneiform tablets, two written in Old-Babylonian Akkadian and two in the Akkadian of the Amarna era. The problem is that traditional chronology places them in the early second millennium BC, even though they belong to a layer corresponding to the peak of the Persian Empire (550–330 BC).

To eliminate the contradiction, historians considered the tablets to be heirlooms, a claim Heinsohn found silly. How can they explain, Heinsohn asked, that the later Hazoreans kept those tablets for more than a thousand years but were incapable of producing some of their own? A similar case comes from the jar with inscriptions in Old-Akkadian (twenty-third century BC) found in the layer of the Hyksos culture, which belonged to the foreign (possibly Asiatic) rulers of Egypt during the Second Intermediate

Period (*c.* seventeenth century BC). Again, historians identified that object as an heirloom.

These absurdities do not validate Fomenko or prove his chronology correct, but they strengthen his arguments against tradition. Still, the question remains: Does this investigation imply that the conventional system is totally wrong?

WHAT YEAR IS THIS, AFTER ALL?

When talking to people about these ideas, I have often been asked what connects the chronology of Europe with that of India, China, or Japan. Because some Far Eastern cultures used a lunar calendar (which makes dating easier) and might have kept better records than the Europeans, their synchronization may solve the problem. The same could be true about the Arab culture, which peaked in medieval times.

Morozov and Fomenko dismissed the Far Eastern chronologies as drawn late, under the influence of the European system. In their view, every Chinese or Korean event that mentioned a contact with the West was dated to fit the traditional view founded by Scaliger and Petavius. An example is found in a Chinese document called the *History Book for Tang Dynasty*, which covers the period 618–907. The text mentions the Roman Empire (大秦) seventeen times. It also tells about a messenger sent by the ruler *Anton* (安敦) to an emperor of the Han dynasty. Historians dated the event to AD 166 and identified *Anton* with Marcus Aurelius Antoninus. But both the meaning of 大秦 as the Roman Empire and the identification of the emperor are far from convincing.

Chronology fares no better with Japan. Part of its history is based on the *Nihon Shoki*, translated as *Nihongi:*

Chronicles of Japan from the Earliest Times to AD 697. Starting where the mythological book *Kojiki* ends, *Nihongi* lists the descendants of the Yamato rulers from the gods, mentions diplomatic contacts with China and Korea, and describes events close to its compilation date of AD 720. But legend and truth are difficult to separate, and the question of its historical significance is similar to the controversy swirling around the Bible.

The chronology of India is in no better shape. In 1965 the renowned Indian historian and Sanskrit scholar Damodar D. Kosambi wrote:

> India has virtually no historical records worth the name . . . In India there is only vague popular tradition, with very little documentation above the level of myth and legend. We cannot reconstruct a complete list of kings . . . What little is left is so nebulous that virtually no dates can be determined for any Indian personality till the Muslim period . . . This has led scholars to state that India has no history.

Even if the chronologies of the Far East were independent and correct, their synchronization with the European system is unclear, so much so that some historians think a global chronology is out of reach. So what can be done to settle the issue?

The radiocarbon checking of basic documents may clarify much about the ancient world. The analysis of Thucydides' papyri, in particular, could fix the dating of the Peloponnesian War. If successful, this project might open the way to a new trend in chronology.

Will historians take up this challenge? If they are confident of their dates, they have nothing to lose. Results that agree with the textbook will strengthen their views and silence the dissent. If the outcome favours Fomenko, things change. Then a strategy for similar projects becomes necessary, and historians would play a central role in designing it.

A failure of tradition would not be a catastrophe—as a friend tried to convince me once. It would mean only that the discipline of history must reconsider its foundations, just as the physical sciences once did. This revision would revive chronology research, which has been marginal in the past century. The present-day feats of science and technology, unimaginable to Scaliger and Petavius, could help us understand many things they didn't.

On a smaller scale, revisionists are already at work in some areas. The foundation of Nicaraguan history, for instance, has been shaken recently. In 2004 a team led by Geoff McCafferty, an archaeologist at the University of Calgary, announced the recovery of 400,000 artifacts from the ancient capital of Quauhcapolca. This find proved that the ancestors of the Nicaraguans didn't come from central Mexico between 1000 and 1300—as previously believed—but had local roots.

Though this example is of no help for the global questions raised by ancient and medieval chronology, it shows how new discoveries can influence our understanding. That doesn't mean that the Thucydides project will affect historians' view. Should its outcome be inconclusive, the problem would remain as puzzling as it is now.

So what year is this, after all? The answer hangs in the balance. Depending on how you weigh the evidence, you might say that mainstream historians know better and that their work proves them right. Or you may think that Isaac Newton, Anatoli Fomenko, and other reformers have a case against tradition and that you are clueless about what year it actually is.

As for me, my doubts about tradition may be superficial or they may be profound. But since I trust the insights of science more than the tales of the past, pursuing the Thucydides project might help me decide. Undoubtedly there will be hurdles, and things can go wrong with fund-raising, collecting samples, or inconclusive results. Yet there is hope.

And if the answer still eludes me, I will have no regrets. For what can inspire more freedom than the quest for a grain of truth?

Notes

Full citations of the references in these notes appear in the References beginning on page 285.

Introduction: Where Did the Time Go?

1 *"Those whose chronology is confused . . .*

Tatian (*c.* 110–180) was an Assyrian philosopher.

7 *the English edition was published*

See Fomenko 1994 for the English version mentioned here.

8 *Then I read the article's title, "Time Warp,"*

The discussion in the second part of this chapter was inspired by "Time Warp," Taylor 2000.

15 *the former world chess champion Garry Kasparov*

Kasparov's ideas about chronology are published in Kasparov 2002.

Chapter 1: Catastrophes and Chaos

21 *"No testimony is sufficient to establish . . .*

Hume 1902, p. 114. As a philosopher and historian, David Hume (1711–1776) was one of the most important figures in the Scottish Enlightenment.

21 *"The Day the Sun Stood Still," by Eric Lallabee*

Though Larrabee's article (Larrabee 1950) was the one that started the debate, John J. O'Neill outlined some of Velikovsky's ideas in 1946 in *The New York Herald Tribune*.

22 *the British Egyptologist Sir Alan Henderson Gardiner*
Sir Alan Henderson Gardiner (1879–1963) was a British Egyptologist. His most important work is the *Egyptian Grammar*, a landmark reference for the Egyptologists of the English-speaking world.

26 *"An extraordinary achievement in a very difficult*
Payne-Gaposchkin 1950b. Cecilia Payne-Gaposchkin published two other articles on *Worlds in Collision*: Payne-Gaposchkin 1950a and 1952.

27 *"The author perhaps does not*
Stewart 1951, p. 61.

27 *Velikovsky responded that he had*
The two articles are Velikovsky 1951a and 1951b.

27 *"I may be all wet*
Velikovsky 1952, Preface.

28 *the chemist Henry Bauer, who made a thorough analysis of this phenomenon*
Henry Bauer's thorough analysis of the Velikovsky phenomenon is published in Bauer 1984.

28 *"I claim the right to fallibility*
Velikovsky 1952, Preface.

28 *"None of my critics*
From the unpublished transcripts of Velikovsky's talk at the San Francisco conference.

28 *In 1955, in a* Scientific American *article*
The *Scientific American* interview with Einstein appeared in Cohen 1955. It followed from a discussion Bernard I. Cohen, a historian of science, had with Einstein while he visited the physicist in his home at 112 Mercer Street in Princeton, New Jersey, two weeks before Einstein's death on April 18, 1955. Velikovsky's name does not appear explicitly in the article, but the context leaves no doubt about the object of Einstein's comment. The quotation "You know, it is not a bad book . . ." is on page 70.

29 *"In the lifetime of Yao*
Velikovsky 1950, p. 101.

30 *In 1883 Ignatius Donnelly published the book*
Ignatius Donnelly (1831–1901) was an American writer, lawyer, and politician who stirred people's imagination with his writings about Atlantis. The book mentioned in the text is Donnelly 1883.

30 *An even more obvious example of how Velikovsky chose to "prove" his point is his mention of Herodotus*
Velikovsky mentions Herodotus in connection with the Sun's motion in Velikovsky 1950, p. 105.

31 *"Egypt was unaffected by this*
Herodotus 1904, Book II, 142.

33 *Another weak link in Stewart's argument was related to the Titius–Bode Law*
The Titius–Bode Law is given by the formula $x = (n + 4)/10$, where n takes the values 0, 3, 6, 12, 24, 48, 96, and 192, for Mercury, Venus, Earth, Mars, the asteroid belt, Jupiter, Saturn, and Uranus, respectively. The value of x is expressed in astronomical units (AU), where 1 AU = 149,597,870.691 kilometres, approximately the distance from Earth to Sun. The results of the computations are given below. Note that the law fails for Neptune and Pluto. In fact, it would work for Pluto if Neptune were missing.

The Titius–Bode Law of Planetary Distances

Planet	*n*	Titius–Bode Law in AU	Approximate Distance in AU
Mercury	0	0.40	0.39
Venus	3	0.70	0.72
Earth	6	1.00	1.00
Mars	12	1.60	1.52
Asteroid belt	24	2.80	2.80
Jupiter	48	5.20	5.20

Saturn	96	10.0	9.54
Uranus	192	19.6	19.2
Neptune	–	–	30.1
Pluto	384	38.8	39.4

34 *The case of two celestial bodies . . . is relatively easy to solve*
The question of who first solved the two-body problem is still debated among historians of science. For a history of various early solutions, see Alain Albouy's article in Cabral and Diacu 2003.

36 *I could evaluate the probability of Velikovsky's collision scenario, and I found it to be extremely small*
The high improbability of a collision between two celestial objects follows from the theorems proved by Donald Saari in Saari 1971 and 1975. He showed that for point masses in a Newtonian gravitational system, the set of initial conditions leading to collisions has zero measure and is of the First Baire Category (a countable union of nowhere dense sets). Since planets are not point masses, it follows that collisions have positive measure. However, this set is so small that collisions are highly improbable.

37 *Laskar had been asked to write a review for the October 2003 issue of* La Recherche
See Laskar 2003.

37 *There is a small probability that Mercury and Venus will come close to each other*
Numerical computations for determining planetary motion have been done since the 1960s, with less accuracy than those made today. Contributions like those of Jacques Laskar, Scott Tremaine, Martin Duncan, Gerald Sussman, Jack Wisdom, and others have improved these methods considerably.

38 *Most of them agreed within 1 arc second*
An arc degree (or a degree of arc) is the angle at which 1/360 of a circle is seen from its centre. An arc minute corresponds to 1/60 of an arc degree, or 1/21,600 of a circle. An arc second corresponds to 1/60 of an arc minute, or 1/1,296,000 of a circle. The diameter of the Moon is

approximately 30 arc minutes. Normally, the naked eye cannot distinguish between two points that are less than 1 arc minute apart.

38 *In 1898 a German schoolteacher, Paul Gerber, gave an explanation for Mercury's motion within the framework of classical mechanics*
See Gerber 1898 and 1917.

39 *In the 1920s the Bulgarian physicist Georgi Manev suggested another model,*
In the summer of 1995 I attended a conference in Plovdiv, Bulgaria, and asked several Bulgarian physicists and mathematicians about Manev. Nobody seemed to have heard of him. In May 2003, after many articles on Manev's problem had been published, the Bulgarian Academy organized a conference in his honour.

Chapter 2: A New Science

42 *"Time is the proper dimension*
Bickerman 1980, p. 9. This is one of the most authoritative books on chronology written in the second part of the twentieth century. Elias Joseph Bickerman (1897–1981) was born in Chisinau, Moldova, and educated in St. Petersburg. In 1918 he reached Germany, where he studied and then taught at the University of Berlin. In 1932, he moved to France to teach at higher institutions, before escaping to the United States after the Nazi occupation. In New York, Bickerman was appointed to the New School for Social Research, then the Jewish Theological Seminary, and from 1952 until his retirement in 1967 he was a professor at Columbia University.

43 *The man who would become the founder of the new science*
Most of the information about Scaliger is taken from Grafton 1983, 1993, and 2003, and Bernays 1965. See Grafton 1975 for an excellent argument that the study of historical chronology reached its climax at the time of Scaliger, and that the field has been in decline since then. More information about the struggle to decipher calendars can be found in Duncan 1998.

43 *"Every day he required from me*
Scaliger 1927, p. 30.

44 *"If one computes backwards*
This passage, from Julius Africanus's *Chronologia*, appears in a slightly different translation in Repcheck 2003.

45 *"For Eusebius and all future chronologists*
Repcheck 2003, p. 31.

46 *"In classical antiquity there was virtually*
Hay 1977, p. 5.

46 *In 1568 Gerardus Mercator*
More information about Mercator's chronology work appears in Crane 2003.

46 *"I do not see how the month of April*
Grafton 1993, p. 37.

48 *An especially useful book for Scaliger was* Liber de epochis
See Crusius 1578.

50 *the respected chronologist Sethus Calvisius*
Sethus Calvisius [Seth Kalwitz] (1556–1615) was a German polymath, born into a peasant family at Gorschleben, Thuringia, on February 21, 1556. His musical talents earned him the directorship of the music school first in Pforten and then in Leipzig. He also received offers for mathematics professorships in Frankfurt and in Wittenberg. In his *Opus chronologicum*, published in Leipzig in 1605, he expounded a system based on the records of nearly three hundred eclipses. He made a proposal for an ingenious reform of the calendar in *Elenchus calendarii Gregoriani*, which appeared in Frankfurt in 1612, and he also wrote several books on music.

50 *the respected astronomer Johannes Kepler*
Johannes Kepler (1571–1630), the famous astronomer who discovered the laws of planetary motion, published the first work on chronology in Linz, in German, in 1613 and the following year he expanded it in Latin under the title *De Vero Anno quo Aeternus Dei Filius Humanam Naturam in Utero Benedictae Virginis Mariae Assumpsit* (Concerning the True Year in which the Son of God Assumed a Human Nature in the Uterus of the Blessed Virgin

Mary). Kepler argued that the Christian calendar was in error by five years and that Jesus had been born in 4 BC.

50 *spring each year arrives about twenty minutes earlier*
This twenty-minute difference occurs because the "sidereal year" (the time the Earth needs to return to its previous position in its motion around the Sun) is about that much longer than the "tropical year" (the time between two spring equinoxes).

51 *"When Zeus has finished sixty wintry days*
Hesiod 8 BC, line 564.

52 *a fact proved only in 1837 by Pierre Wantzel*
Wantzel 1837. Charles Sturm later provided simpler proofs, but he didn't publish them.

52 *"The few modern historians who mentioned Scaliger*
Grafton 2002, p. 4. Grafton's assessment is corroborated by other sources. For example, Wachsmuth 1895, p. 10, stated: "[Joseph Scaliger] was the founder of technical chronology, in both the theoretical and technical sense and in its practical application to history." Similarly, Ginzel 1906, p. 54, wrote: "Joseph Scaliger . . . should be seen as the founder of technical chronology." In addition to these views, see Grafton 2003, pp. 219–20: "In an age of encyclopaedias and encyclopaedic claims, of disciplines that claimed to offer the keys to all the kingdoms of knowledge, chronology covered a wider range than most, and offered perhaps more opportunities than any other field for the display of extravagant erudition and divinatory virtuosity. No wonder that Scaliger became hooked on the challenge of this uniquely complex field, or that this lonely genius managed to revolutionize it by transforming cosmic portents into chronological benchmarks."

53 *The amendment of the parallel's postulate*
The parallel's postulate of Euclidean geometry states that, through a point that does not lie on a given line, one can draw a single parallel to that line. In the nineteenth century, the mathematicians János Bolyai, Nikolai Lobachevski, and Bernhard Riemann replaced this axiom with different ones, thus constructing the first non-Euclidean geometries.

53 *Among the new critics was Denys Pétau*
Most of the information on Petavius is taken from Di Rosa 1960.

54 *Petavius used the combined cycles method extensively*
The combined cycles method used by Scaliger and Petavius has its origins in the twelfth century. For historical details and the mathematical reasons why 4713 BC was taken as the base date for the Julian count, see Reese 1981 and 1983. Though Petavius did not use the Julian count in his computations, the idea of the method is easier to convey in these terms. Nevertheless, my explanation doesn't change the essence of the method.

54 *The combined cycles method assigns to every date in history its corresponding Julian count*
The calendrical computation of dates uses the Julian calendar. The first Julian day (called, in fact, Julian day zero) is noon on Monday, January 1, 4713 BC. In the Gregorian calendar, this date corresponds to noon on Monday, November 24, 4714 BC (see Reid 2001, p. 18, and for the negative-year notation used there, see p. 15).

55 *The French Jesuit Jean Hardouin*
For more about Jean Hardouin (1646–1729), see Mencken 1937.

56 *His more famous brother, Theodor . . . criticized August in an 1858 book*
Mommsen 1858. Later, his brother August published his research in book form (see Mommsen 1883).

56 *In the second part of the nineteenth century*
A notable predecessor of Ginzel and von Oppolzer was Heinrich Bünting, see Bünting 1590. Bünting was one of the first chronologists who emphasized the importance of using the information on the time intervals between astronomical events for historical dating (see Grafton 2003, p. 215). An important follower of the Austrian astronomers was Justin Schove. A close investigation of his book, Schove 1984, makes it clear that only a very few ancient and astronomical records corroborate the historical dates of traditional chronology.

56 *the Austrian astronomers Friedrich Ginzel and Theodor von Oppolzer sought to set traditional chronology on a firmer base*
The works of Ginzel and von Oppolzer that meant to put traditional chronology on a firm foundation are Ginzel 1899, 1906/1911/1914, and 1908, and Oppolzer 1962.

58 *During this time, Morozov wrote poetry*
Bullock 1992, p. 32, explains that, in the Russian revolutionary tradition, prison served as a "university." In prison, inmates read widely, discussed ideas, and were instructed by their more educated comrades. He shows that, in spite of being imprisoned for more than twenty-five years, Morozov lived in an intellectual environment that stimulated his unusual capacity for creativity and learning.

59 *The fruit of his research was the seven-volume treatise* Christ
See Morozov 1997/98 for a recent edition of Nicolai Morozov's book.

60 *But this time he approached Felix Dzerzhinsky*
Felix Dzerzhinsky (1877–1926), a Pole, was also a founder of the Social Democratic Polish Party. In 1923, when Morozov approached him, he was people's commissar for transport as well. In 1922 he transformed the Cheka into the State Political Administration (GPU), which had the same role of silencing any opposition to Lenin and, later, to Stalin. Between 1919 and 1924 he ordered the execution of at least 200,000 people. See Bullock 1992, p. 61.

61 *Morozov considered that the mistake had been made in a misapprehension of the Saturn–Jupiter cycle*
The sixty-year Saturn–Jupiter cycle is only an approximation. In fact, Jupiter's period is 11.88 years, whereas Saturn's is 29.42 years, so 11.88 x 5 = 59.4, and 29.42 x 2 = 58.84.

62 *"The cataclysms of Chinese history*
Butterfield 1981, p. 143.

63 *"I had to address several distinguished historians*
Fomenko 2003a, p. xxii.

65 Fomenko's chronology is summarized in figure 2.4

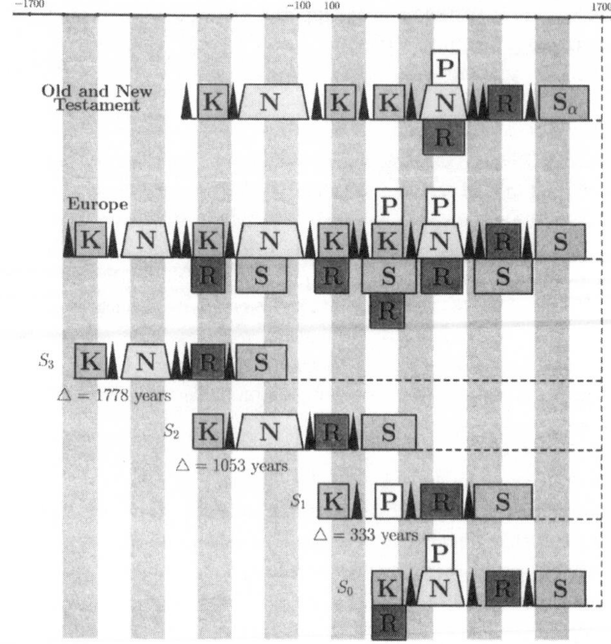

Figure 2.4—Fomenko's three shifts, of 333, 1053, and 1778 years. Fomenko claims that, in a first approximation, an original chronicle was repeated three times. In fact, the sequences in each shift are not identical but resemble each other. S0 is the original chronicle, and S_1, S_2, S_3 are the shifts. Every block in each sequence represents a repetition. For example, the block K describes the same events; the same holds for the black triangle. The top sequence summarizes Fomenko's view on the repetitions in the Bible, and the one below it refers to Europe. In both cases some blocks overlap, which means that the historical events they represent are intermingled.

Chapter 3: Swan Song

67 *"Nature and Nature's laws*
Alexander Pope (1688–1744), "Epitaph: Intended for Sir Isaac Newton" (1730).

67 *Some biographers*
The biographies of Isaac Newton mentioned here are Westfall 1980, Clark 2001, and Gleick 2003. For a less dense biography of Newton, see Westfall 1993.

70 *"Sir, six months ago*
Newton MS (K), Keynes Manuscript 138, letter of Guillaume Cavelier to Newton, March 20, 1725.

72 *"[In the year] 939, the ship Argo*
Newton 1728a.

74 *"For Hipparchus tells us*
Newton 1728b, pp. 82–83.

77 *Chiron had an interest in helping the expedition*
The two grandchildren of Chiron who took part in the Argonautic Expedition were Peleus and Telamon.

78 *"Let him have but a stone doll*
Spence 1820, p. 25.

79 *"Does one have to conclude from it*
Fréret 1758, p. 418.

79 *"clearly explained . . ."* and *"worthy to have statues*
Reid 1728, p. 349.

82 *"a frivolous system . . . imaginary and chimerical."*
Hardouin 1729, p. 1582.

83 *"Fréret disputed most of Newton's textual interpretation*
Manuel 1963, p. 183.

84 *"The name of Newton raises the image*
British Museum, London, Manuscript 34880, Edward Gibbon, *Commonplace Book.*

84 *"Already full of esteem for this man of letters*
Gibbon 1761, p. 151.

87 *"To show that the Israelites rather than the heathen*
Manuel 1963, p. 193.

88 *In their 1991 book* Centuries of Darkness
See James 1991.

88 *the British historian Sir Flinders Petrie*
Sir Flinders Petrie (1853–1942) is considered the father of modern archaeology. Legend has it that while surveying the Great Pyramids of Egypt, he walked naked or wore only a ballerina's tutu—because the locals would beat or kill anyone suspected of ransacking the pyramids but left the insane alone.

89 *To historians, Kitchen's 1973 book marks the chronological foundation*
See Kitchen 1983, the book that laid the foundations for the chronology of Egypt's Third Intermediate Period.

89 *In a review published on May 17, 1991*
Kitchen's review of *Centuries of Darkness* as well as the three mentioned letters appeared in *The Times Literary Supplement*: the review on May 17, 1991, p. 21; James's first letter on June 7, 1991, p. 15; Kitchen's response on June 21, 1991, p. 13; and James's second letter on July 12, 1991, p. 13.

90 *"In the course of a single century's research*
Lehmann 1977, p. 204.

91 *His first book,* A Test of Time
See Rohl 1995.

92 *1517 BC, mentioned in a medicine text called the Ebers Papyrus*
Georg Moritz Ebers (1837–1898) was born in Berlin, studied law and then hieroglyphs, and was professor of Egyptology at the

University of Leipzig from 1875 to 1889. He wrote numerous scientific books on Old Egypt, as well as several novels. The Ebers Papyrus was discovered about 1862 and purchased by Ebers in Luxor, Egypt, in 1873.

93 *Kitchen invoked another inscription, this one on a statue in the British Museum*
The statue in the British Museum to which Kenneth Kitchen refers when invoking the marriage between the son of Shoshenk I and the daughter of Psusennes II has the code BM8.

Chapter 4: Historical Eclipses
Information about all historical eclipses, including quotes from the original texts that describe them, can be found at http://hbar.phys.msu.su/gorm/atext/ginzele.htm.

97 *"Chronology is nothing but the computation*
Grafton 2003, p. 218. A note about Sethus Calvisius appears in this book on page 264.

99 *The Peloponnesian War between the Greek city-states of Athens and Sparta*
Strassler 1996 provides a comprehensive guide to the Peloponnesian War.

99 *"The same summer, at the beginning*
Thucydides 1849, 2.28.1.

99 *"In the first days of the next summer*
Thucydides 1849, 4.52.1.

100 *"All was ready, and they*
Thucydides 1848, 7.50.4.

100 *In* De emendatione temporum, *Scaliger provides dates*
Ancient Greeks divided the year into summer and winter, the former season extending from March to November. That two-way division explains why an eclipse in March is considered to have taken place in the summer.

100 *"These things are so well determined*
Newton MS (NC), II, fol. 114 recto and verso.

100 *More research followed, and some astronomers redid the calculations*
Several calculations have been made about the 431 BC eclipse in Athens, all of which are mentioned in Fomenko 1994, vol. 1. The following results were obtained by various authors, depending on what corrections were taken into account; the percentage indicates how much of the Sun's disk was covered by the Moon at the maximum phase of the eclipse: Petavius 85.4%, Struyck 91.6%, Zech 86.5%, Hoffman 89.33%, and Heis 65.83%. However, the recent computations provided by Stephenson and Fatoohi are the most reliable.

101 *In his* Life of Pericles, *the Roman biographer Plutarch*
See Plutarch 1928, "Life of Pericles," 35.1–35.2, for his description of the eclipse.

101 *The Roman orator Cicero reported the same eclipse*
See Cicero 1928, 1.16.25, for his mentioning of the eclipse.

101 *Mars stood only 3 degrees of arc above the line of the horizon*
There are two reasons why Mars, being only 3 degrees above the horizon, was unlikely (but not necessarily impossible) to be seen during the 431 BC eclipse: first, even a slight deformation of the landscape is higher than 3 degrees (about 6 Moon diameters); and, second, during total eclipses, the horizon may not get dark if the Sun is high in the sky.

101 *The latest word on the 431 BC eclipse has come from two British researchers*
See Stephenson and Fatoohi 2001.

104 *Fomenko checked the translation of the original document with the linguist E. V. Alexeeva*
The text analysis of E.V. Alexeeva appears in Fomenko 2003a, pp. 471–73.

104 *Thucydides wrote that, during the war, "eclipses of the Sun occurred with a frequency unrecorded in previous history."*
Regarding the frequency of eclipses visible in the Peloponnese, the ones that took place in the time of the conventional solutions were in 431, 426, 424, 418, 411, 409, 405, and 404, and sixteen half a century earlier: in 480, 478, 477, 470, July and December 466, 463, 458, 455, 453, 450, 448, 447, 437, 434, and 433. From AD 1133 to 1160 the seven eclipses visible from the Peloponnese were in 1133, 1137, 1138, 1140, 1141, 1147, and 1153. The fifteen eclipses in the previous fifty years took place in 1084, 1086, 1087, 1091, 1093, 1098, 1106, 1109, 1112, 1113, 1115, 1119, 1124, 1125, and 1131. Again, the frequency is much the same during both periods. In Fomenko's case, between AD 1039 and 1066, there were eleven eclipses: in 1039, 1044, 1046, 1047, 1053, 1055, 1058, 1059, 1064, 1065, and 1066, while, during the preceding fifty years, only fifteen: in 990, 992, 993, 999, 1000, 1004, 1006, 1007, 1010, 1015, 1017, 1021, 1030, 1032, and 1033.

105 *the war took place during a period of disasters that had no previous match in history*
The passage on "unparalleled extent and violence" appears in Thucydides 1859, 1.23.3.

106 *But in the original Greek the word Thucydides wrote is mnemoneuomena*
On October 1, 2003, a personal communication with Mortimer Chambers, professor of history at UCLA, an expert on Thucydides as well as on ancient Greek and Roman history, confirmed, first, that Thucydides is logical and trustworthy in his book; but, second, that there seems to be one exception: the passage in which he mentions the high frequency of eclipses during the Peloponnesian War; that is because he associates the eclipses with unusual catastrophes; and, third, that the word "recorded" is the translation of the Greek participle *mnemoneuomena*, "remembered"—"in people's memories"—a term that has no reference to the methodical recording or cataloguing of eclipses.

107 *"When the consul [Publius Africanus] left for the war*
Livy 1890, Book 37, 4.4.

108 *"When it was night and, supper*
Plutarch 1928, "Life of Aemilius Paulus," 25.1.

109 *From this and other contextual information given by Livy and Plutarch*
Livy's description of the eclipse appears in Livy 1890, Book 44, 36.1, "It was past the summer solstice and the time of day was approaching noon . . . ," and 37.8, "When the fortification of the camp was completed, C. Sulpicius Gallus, a military tribune attached to the second legion, who had been a praetor the year before, obtained the consul's permission to call the soldiers on parade. He then explained that on the following night the Moon would lose her light from the second hour to the fourth, and no one must regard this as a portent, because this happened in the natural order of things at stated intervals, and could be known beforehand and predicted. Just in the same way, then, as they did not regard the regular rising and setting of the Sun and Moon or the changes in the light of the Moon from full circle to a thin and waning crescent as a marvel, so they ought not to take its obscuration when it is hidden in the shadow of the Earth for a supernatural portent. On the next night [September 4] the eclipse took place at the stated hour, and the Roman soldiers thought that Gallus possessed almost divine wisdom. It shocked the Macedonians as portending the fall of their kingdom and the ruin of their nation, nor could their soothsayers give any other explanation. Shouts and howls went on in the Macedonian camp until the moon emerged and gave her light." This eclipse was also mentioned by Polibius, Cicero, Valerius Maximus, Frontinus, Pliny, and Quintilianus (see http://hbar.phys.msu.su/gorm/atext/ginzele.htm). According to tradition, none of those who described this eclipse witnessed it, since they all lived more than a century and a half after it happened.

109 *Those from 955 and 1020 had the longest duration*
The phase of the AD 955 lunar eclipse was 16"1, and the one for the AD 1020 lunar eclipse was 18"7. Note that 12" is already total, and that the maximum possible phase is 22"7. Fomenko 1994, vol. I, p. 24.

110 *Nicolai Morozov pointed out the existence of a long-lasting lunar eclipse*
The phase of the AD March 21, 368, lunar eclipse was 13"3. Fomenko 1994, vol. I, p. 25.

111 Calendar Reform and the Council of Nicaea
See Nosovski 1994 and 2003 for the topics discussed in this section.

111 "If you want to find out which year it is
See *Liber de Paschate* at http://hbar.phys.msu.ru/gorm/chrono/paschata.htm. The original Latin text reads: "Si nosse vis quotus sit annus ab incarnatione Domini nostri Jesu Christi, computa quindecies XXXIV, fiunt DX; iis semper adde XII regulares, fiunt DXXII; adde etiam indictionem anni cujus volueris, ut puta, tertiam, consulatu Probi junioris, fiunt simul anni DXXV. Isti sunt anni ab incarnatione Domini."

111 nobody disputes that the division of time beginning with AD 1 came into existence with this manuscript
More information about the origins of the Christian era appear in Declercq 2000.

112 There are documents that connect Jesus to Tiberius
The quotation connecting Jesus with Tiberius appears in Tacitus 1907, 15.44. Pilate was, in fact, a prefect, not a procurator, so Tacitus is mistaken about the title.

113 "Our care was not only to
Fomenko 1994, vol. II, p. 398.

116 nine others fulfilled the basic requirements
The nine other eclipses alleged to have taken place between 100 BC and AD 1700 occurred in AD 1190, 848, 753, 658, 574, 479, 137, and 53 and in 43 BC.

118 But the dating of the 1054 supernova has been disputed
See Lupoato 1997 for a discussion of these claims.

Chapter 5: The Moon and the Almagest

120 "When I follow the windings of heavenly bodies
This sentence appears in Book I of the *Almagest*, following the table of contents. Claudius Ptolemy (*c.* AD 85–165) did his work in Alexandria, Egypt. His nationality, place of birth, and place of death are unknown.

121 *Fomenko looked at what researchers call the Moon's elongation*
The elongation of the Moon is defined as the difference in longitude between the mean Moon and the mean Sun. These terms refer to the Moon and the Sun as if they moved uniformly, like the hands of a clock. The mean Moon and the mean Sun move close to the real Moon and the real Sun but do not overlap with them.

121 *the Moon's elongation, denoted by the expression D"*
See Fomenko 1981 on D."

122 *In 1979 Robert R. Newton, a professor at Johns Hopkins University*
See Newton 1979 and 1984 for a discussion of the Moon's acceleration.

124 *"We have found too many instances*
Newton 1984, p. 186.

125 *"Either the scarce astronomical descriptions*
Fomenko 2003a, p. 105.

126 *Ptolemy wrote the* Almagest *during the reign of the Roman emperor Antoninus Pius*
Pappus (*c*. 320) and Theon of Alexandria (*c.* 370) offered commentaries on the original version of the *Almagest*, which is lost. Arab translations appeared as early as the ninth century under the title "Al-majasti," derived from the Greek word *megiste* (μεγιστε), meaning "greatest treatise." A Greek copy arrived in Rome from Byzantium around 1450. Ten years later the Cretan philosopher Georgius Trapezuntius made the first Latin translation. An abbreviated version was published in 1496 under the supervision of the mathematician Johann Müller, better known as Regiomontanus, and an edition containing celestial maps by Albrecht Dürer appeared in Venice in 1515. The *Almagest* has been published in dozens of languages all over the world.

126 *In 1977 Robert Newton published* The Crime of Claudius Ptolemy
See Newton 1977 on Ptolemy's alleged fabrications.

127 *"Did Ptolemy do any observing?*
Delambre 1817, vol. 2, p. xxv.

127 *"Delambre and Newton have convincingly proved*
Waerden 1988, p. 253.

128 *Gerd Graßhoff, from the University of Bern*
See Graßhoff 1990 for his assessment.

128 *Oscar Sheynin of Berlin*
See Sheynin 1993 for his opinion.

128 *In the opinion of James Evans of the University of Puget Sound*
See Evans 1993 for his opinion.

128 *But why regard Newton's work as either black or white?*
A description of the pros and cons of Robert Newton's ideas regarding the falsification of Ptolemy's observations appears in Jonsson 1986.

128 *In the 1920s Morozov had already thought about the* Almagest
See Morozov 1997/98 for his work on Ptolemy's star catalogue.

131 *he also thought something was wrong with the dating of the star catalogue*
See Fomenko 1989 for his work on the dating of Ptolemy's star catalogue, as well as Fomenko 1994, vol. II, Appendix 2, pp. 346–75.

131 *The best version turned out to be the one by Christian Peters and Edward Knobel*
Peters and Knobel 1915 was the book Fomenko used to date Ptolemy's catalogue of stars.

137 *The Russian mathematicians didn't stop here*
The works of Fomenko, Kalashnikov, and Nosovski about the dating of Ptolemy's star catalogue, occultations, and eclipses are Fomenko 1989, 1992, and 1993.

137 *The* Almagest *mentions four occultations*
Fomenko's study on the dating of the four occultations and the eighteen lunar eclipses recorded in the *Almagest* appears in Fomenko 1992 as well as in Fomenko 1994, vol. II, Appendix 3, pp. 376–89.

137 "We again took one of the precisely
A similar translation appears in Toomer 1984, p. 522.

138 the eras of Nabonassar, Dionysius, and "after the death of Alexander,"
Alan Samuel attributes the dating by "years from the death of Alexander" to the era of Philip Arrhidaios, after a king of Macedonia who reigned between 323 and 317 BC. For more information on the eras of Arrhidaios, Nabonassar, and Dionysius, see Samuel 1972, pp. 51–52.

139 Ptolemy wrote that Venus "occulted" η-Virgo
In the *Almagest*, for Venus "occulted" η-Virgo, Ptolemy uses the Greek word κατειληφωζ, and for Jupiter "covered" δ-Cancer, he employs επεκαλνψεν.

Chapter 6: Ancient Kingdoms
Most information in this chapter is taken from Fomenko 2003b.

145 there are 3,732,480 possible configurations of these heavenly bodies
Except for Mercury, whose angular distance from the Sun cannot exceed 28 degrees of arc, and for Venus, 48 degrees, the other planets may show up anywhere. Since every constellation spans 30 degrees, for a fixed position of the Sun, Mercury can occur in three constellations, Venus in five, and the other planets in all twelve. Consequently there are 3 x 5 x 12 x 12 x 12 x 12 x 12 = 3,732,480 possible horoscopes.

147 "If this were only
Morozov 1997/98, vol. 6, p. 653.

149 In 1883 he interpreted the figures
The identifications appear in Brugsch 1968, volume 1, first published in 1883.

149 In 1977 the French Egyptologist Sylvie Cauville published a new study
See Cauville 1977.

150 *But, in the 1990s, the Russian physicists N.S. Kellin and D.V. Denisenko proved*
When computing Morozov's solution for the long Denderah zodiac, Kellin and Denisenko showed that, on AD May 6, 540, Mercury was between 15 and 17 arc degrees east of the Sun, which made the planet visible in Moscow and the more so in Egypt.

151 *he had relied on a drawing of the long zodiac published in 1802*
See Denon 1802 and 1986.

152 *The smaller figures resembling large characters must be planets indicating sky configurations at solstices and equinoxes*
For the long Denderah zodiac, the fall equinox falls correctly during a configuration seen between September 5 and 18, 1167. The partial horoscope indicates Venus in Virgo and the new moon near Venus. The winter solstice, December 5–18, 1167, is indicated by Mercury in Sagittarius, Venus and Saturn in Capricorn, and Mars on the boundary between Scorpio and Libra. The spring equinox, March 7–20, 1167, is indicated by Jupiter in Pisces and the summer solstice, June 6–18, 1168, by Venus and Mercury in Taurus.

156 *The first person to attempt a dating of the zodiacs was the English astronomer Edward Knobel*
Edward Ball Knobel (1841–1930) was a British businessman and amateur astronomer. His work on a publication about the chronology of star catalogues in 1875 led him to the study of the work of early Arab astronomers.

156 *"The year 59 AD, January*
Morozov 1997/98, vol. 6, p. 732, taken from a report published in 1908. British School of Archaeology in Egypt and Egyptian Research Account.

159 *another painted on a wall in the Petosiris tomb of Dakhla*
When Neugebauer, Parker, and Pingree (1982) researched the Petosiris zodiac, they concluded that the mural was influenced by Mithraism, a pre-Christian religion.

161 *In 1959 Otto Neugebauer and H.B. Van Hosen published a study*
See Neugebauer and Van Hosen 1959.

Chapter 7: Overlapping Dynasties

163 "Time present and time past
Eliot 1944. T.S. Eliot, poet, 1888–1965.

164 "[Kolmogorov] said he was frightened
Fomenko 2003a, p. xxxii.

171 Other popes are included in Liber Pontificalis
Loomis 1916 is one English translation of the Book of Popes. More information about the popes can be found in Schimmelpfennig 1992.

Chapter 8: Secrets and Lies

187 "There are three kinds of lies
Attributed to the British novelist and politician Benjamin Disraeli (1804–81) in Twain 1961.

198 "The scientific results obtained by the author
Fomenko 2003a, p. xix.

199 "The whole thing is as nutty as can be
Owen Gingerich's statement "The whole thing is as nutty as can be . . . " appeared in Beam 1991.

199 A similar opinion comes from the Russian mathematician Sergei P. Novikov
See Novikov 2000.

199 "[Aivazyan] had already considered
Novikov 2000, p. 366.

204 "In arithmetic it would look like this
Zalyzniak 2000, p. 382.

205 Zalyzniak couldn't find a single correct interpretation
Zalyzniak has many more arguments against the etymological practices of Nosovski and Fomenko. For example, he shows that they identified *Rus* (Russia) and *russkij* (Russian) with *ulus* (region of the Golden Horde), *Rosh* (the name of a biblical country),

Irish (from Ireland), *Ross* (horse in German), and the geographical names *Prussia*, *Saraj*, *Saranks*, and *Saratov*. But based on this "principle," Zalyzniak claims, they could have made many more associations, and he provided a few: *rus* (village in Latin), *Rusa* (town in Russia), *Ruza* (river in Russia), *rusalka* (mermaid), *rys* (lynx), *russus* (red in Latin), *rosse* (jade in French), *ours* (bear in French), *rosvo* (robber in Finnish), *Ruslan* (Russian name), *Rousseau* (French name), *surovyj* (severe), *sor* (litter), *Sura* (river in Russia), *Saar* (river in Europe), *Syria*, *zulus* (South African tribe), *G-Ruzia* (Georgia), *Pe-Rsia*, *Je-Rusalem*, *tu-rusy* (idle gossip), etc.

208 "We believe that the unprejudiced reader
This and the next three translations in the paragraph for Fomenko's quotations appear in Zalyzniak 2000, p. 402.

208 "this is the position of a prophet
Zalyzniak 2000, p. 403.

208 "Fomenko's doctrine in its present form
Zalyzniak 2000, p. 403.

209 "South America was initially inhabited
This and the other two quotations in the paragraph appear in Zalyzniak 2000, p. 404.

Chapter 9: Scientific Dating

213 "Everything that has come down to us
See http://www.grahamhancock.com/forum/HancockS1-p1.htm. Rasmus Nyerup (1759–1829) was a Danish historian and librarian.

214 Radiocarbon Dating
Some general information about scientific dating techniques has been taken from Fleming 1976.

214 Libby presented a new theory for dating objects
A history of Libby's invention of the radiocarbon method appears in Taylor 1987, chapter 6.

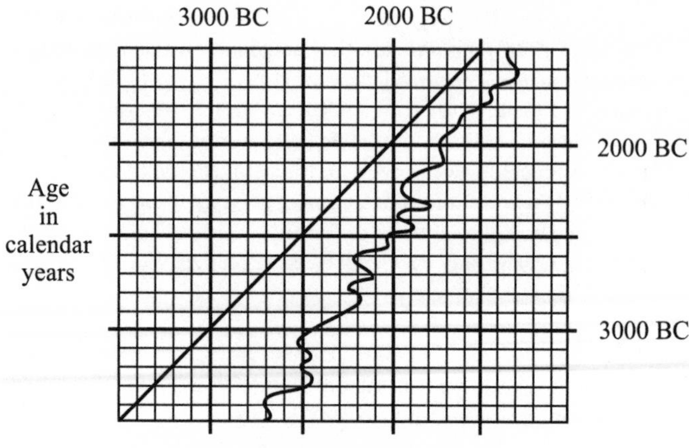

Figure 9.3—This graph presents an early state of the calibration method. The horizontal axis denotes radiocarbon years and the vertical one, calendar years. If, say, the radiocarbon date of a sample reads 2000 BC, the correction curve indicates the calendar date of 2500 BC. The oblique line shows that, in the time interval of the diagram, uncalibrated radiocarbon dates would be younger than tree rings suggest. The main problem, however, is that the kinks of the curve give sometimes multiple dates for a single reading.

218 Today, radiocarbon results are calibrated with the help of other disciplines
An early calibration of the radiocarbon method using dendrochronology is represented in figure 9.3.

221 "The Sun became dark and its darkness lasted
This quotation appears in the *Chronicles of Michael the Syrian*, 9.296.

222 a group of geologists from Columbia University studied corals
See Bard et al. 1990 for the study of the corals raised off the island of Barbados.

225 *Another method used for the time range of history is archeomagnetic dating*
Most of the information about archeomagnetic dating comes from Eighmy and Sternberg 1990.

228 *"[The laboratories] received specimens of wood*
Fomenko 2003a, p. 79.

229 *"The margin of error*
Coghlan 1989, p. 26.

230 *In 1995 a team of experts from the University of Arizona in Tucson analyzed the Dead Sea Scrolls*
The dating of the Dead Sea Scrolls is detailed in Bonani et al. 1992 and in Erikson 1995.

Chapter 10: Finding a Consensus

235 *"I'm sorry I can't give you*
From the author's private email correspondence with Gregory Rowe.

238 *"Ethiopians make their gods snub-nosed*
This quotation is an epigraph in Toynbee 1934 and is taken from Diehl 1922, pp. 58–59.

240 *"we have found too many instances*
Newton 1984, p. 186.

242 *"The Bible is right after all!*
Keller 1995, p. 24.

243 *Harpur believes that Jesus didn't exist*
See Harpur 2004 for these ideas.

243 *Wilson was inclined to agree with Keller*
See Wilson 2000.

243 *The two experts think the Bible is pure mythology*
See Finkelstein and Silberman 2001.

246 "The work discussed here
Fomenko 1994, vol. I, p. 13.

247 "We believe that the unprejudiced reader
This and the next three translations in the paragraph of Fomenko's quotations appear in Zalyzniak 2000, p. 402.

247 "ideas are perfectly rational
See Fomenko 2003a, p. xix, for Shyraev's quotation.

247 "exceptional scientific scrupulousness
This quote appears in Zinoviev's assessment of Fomenko's work in Fomenko 2003a, pp. xv–xvii.

250 "The good Christian should beware
Saint Augustine's quotation, freely translated, is commonly found in the mathematical folklore. It appears in *De Genesi ad Litteram Libri Duodecim* (The Literal Meaning of Genesis in Twelve Books), vol. 2, chapter 17: "Bono christiano sive mathematici sive quilibet inpie diviantium, maxime dicentes vera, cavendi sunt, ne consortio daemoniorum animam deceptam pacto quodam societatis inretiant."

254 Part of its history is based on the Nihon Shoki, *translated as* Nihongi
See Nihongi 1956 for this reference.

255 "India has virtually no historical records
Kosambi 1965, pp. 9–10.

256 In 2004 a team led by Geoff McCafferty
An announcement about McCafferty's work on the foundation of Nicaraguan history appeared in Walton 2004.

References

This selected list provides complete references for the publications cited by author and date in the Notes section.

Bard et al. 1990
Bard, E., B. Hamelin, R.G. Fairbanks, and A. Zinder. "Calibration of the 14C Timescale over the Past 30,000 Years Using Mass Spectrometric U-Th Ages from Barbados Corals." *Nature* 345 (1990): 405–10.

Bauer 1984
Bauer, Henry H. *Beyond Velikovsky: The History of a Public Controversy.* Urbana and Chicago: University of Illinois Press, 1984.

Beam 1991
Beam, Alex. "A Shorter History of Civilization." *Boston Globe*, September 16, 1991.

Bernays 1965
Bernays, Jacob. *Joseph Justus Scaliger.* Osnabrück, West Germany: Otto Zeller Verlag, 1965. Reprint of the 1855 edition.

Bickerman 1980
Bickerman, E.J. *Chronology of the Ancient World.* London: Thames and Hudson, 1980.

Bonani et al. 1992
Bonani, G., S. Ivy, W. Wölfli, M. Broshi, I. Carmi, and J. Strugnell. "Radiocarbon Dating of Fourteen Dead Sea Scrolls." *Radiocarbon* 34 (1992): 843–49.

Brugsch 1968
Brugsch, Karl Heinrich. *Thesaurus Inscriptionum Aegypticarum: Altägyptische Inschriften.* Graz, Austria: Akademische Druck u. Verlaganstalt, 1968. Reprint of the 6-volume Leipzig edition, 1883–91.

Bullock 1992
Bullock, Alan. *Hitler and Stalin: Parallel Lives.* New York: Knopf, 1992.

Bünting 1590
Bünting, Heinrich. *Chronologia.* Magdeburg, 1590.

Butterfield 1981
Butterfield, Herbert. *The Origins of History.* London: Eyre Methuen, 1981.

Cabral and Diacu 2003
Cabral, Hildeberto, and Florin Diacu, eds. *Classical and Celestial Mechanics: The Recife Lectures.* Princeton, NJ: Princeton University Press, 2003.

Cauville 1977
Cauville, Sylvie. *Dendera: Les chapelles osiriennes.* Cairo: Institut français d'archéologie orientale du Caire, 1977.

Cicero 1928
Cicero, Marcus Tullius. *The Republic (De republica).* Cambridge, Mass.: Harvard University Press, 1928.

Clark and Clark 2001
Clark, David H., and Stephen P.H. Clark. *Newton's Tyranny: The Suppressed Scientific Discoveries of Stephen Gray and John Flamsteed.* New York: W.H. Freeman, 2001.

Coghlan 1989
Coghlan, Andy. "Unexpected Errors Affect Dating Techniques." *New Scientist* 1684 (30 September 1989): 26.

Cohen 1955
Cohen, Bernard I. "An Interview with Einstein." *Scientific American* 193 (July 1955): 68–73.

Crane 2002
Crane, Nicholas. *Mercator: The Man Who Mapped the Planet.* London: Weidenfeld & Nicolson, 2002.

Crusius 1578
Crusius, Paulus. *Liber de epochis.* Basle, 1578.

Declercq 2000
Declercq, Georges. *Anno Domini: The Origins of the Christian Era.* Turnhout, Belgium: Brepols, 2000.

Delambre 1819
Delambre, Jean Baptiste Joseph. *Histoire de l'astronomie ancienne.* Paris: Chez Mme. Veuve Courcier, 1819.

Denon 1802
Denon, Vivan. *Planches du voyage dans la Basse et la Haute Egypt.* Paris: Didot L'Aîné, 1802.

Denon 1986
Denon, Vivan. *Travels in Upper and Lower Egypt during the Campaigns of General Bonaparte.* 2 vols. London: Darf, 1986.

Diehl 1922
Diehl, E. *Antologia lyrica.* Leipzig, Germany: Teubner, 1922.

Di Rosa 1960
Di Rosa, Pietro. "Denis Petau e la cronologia." *Archivum Historicum Societatis Iesu* 57/ XXIX (1960): 3–54.

Donnelly 1883
Donnelly, Ignatius. *Ragnarok: The Age of Fire and Gravel.* New York: D. Appleton, 1883.

Duncan 1998
Duncan, David Ewing. *Calendar: Humanity's Epic Struggle to Determine a True and Accurate Year.* New York: Avon Books, 1998.

Eighmy and Sternberg 1990
Eighmy, Jeffrey L., and Robert S. Sternberg. *Archeomagnetic Dating.* Tucson: University of Arizona Press, 1990.

Eliot 1944
Eliot, Thomas Stearns. *Four Quartets.* London: Faber and Faber, 1944.

Erikson 1995
Erikson, Jim. "UA Confirms Dead Sea Scrolls Predate Christianity." *Arizona Daily Star*, April 12, 1995.

Evans 1993
Evans, James. "Ptolemy Indicted Again." *Journal for the History of Astronomy* 24 (1993): 145–47.

Finkelstein and Silberman 2001
Finkelstein, Israel, and Neil Asher Silberman. *The Bible Unearthed: Archaeology's New Vision of Ancient Israel and the Origin of Its Sacred Texts.* New York: Free Press, 2001.

Fleming 1976
Fleming, Stuart. *Dating in Archeology: A Guide to Scientific Techniques.* London: J.M. Dent and Sons, 1976.

Fomenko 1981
Fomenko, Anatoli T. "The Jump of the Second Derivative of the Moon's Elongation." *Celestial Mechanics* 29 (1981): 33–40.

Fomenko 1994
Fomenko, Anatoli T. *Empirico-Statistical Analysis of Narrative Material and Its Applications to Historical Dating.* Vol. 1: *The Development of Statistical Tools.* Vol. 2: *The Analysis of Ancient and Medieval Records.* Dordrecht, Netherlands: Kluwer Acad. Publ., 1994.

Fomenko 2003a
Fomenko, Anatoli T. *History: Fiction or Science? Chronology 1.* Isle of Man: Delamere Resources, 2003.

Fomenko 2003b
Fomenko, Anatoli T., Tatiana N. Fomenko, Wieslaw Z. Krawcewicz, and

Gleb V. Nosovski. "Mysteries of Egyptian Zodiacs." Unpublished manuscript, Edmonton, 2003.

Fomenko, Kalashnikov, and Nosovski 1989
Fomenko, Anatoli T., V.V. Kalashnikov, and G.V. Nosovski. "When Was Ptolemy's Star Catalogue in 'Almagest' Compiled in Reality? Statistical Analysis." *Acta Applicandae Mathematicae* 17 (1989): 203–29.

Fomenko, Kalashnikov, and Nosovski 1992
Fomenko, Anatoli T., V.V. Kalashnikov, and G.V. Nosovski. "The Dating of Ptolemy's 'Almagest' Based on the Coverings of the Stars and on Lunar Eclipses." *Acta Applicandae Mathematicae* 29 (1992): 281–98.

Fomenko, Kalashnikov, and Nosovski 1993
Fomenko, Anatoli T., V.V. Kalashnikov, and G.V. Nosovski. *Geometrical and Statistical Methods of Analysis of Star Configurations: Dating Ptolemy's Almagest.* Boca Raton, Fl.: CRC Press, 1993.

Fréret 1758
Fréret, Nicolas. *Défense de la chronologie.* Paris: Bougainville, 1758.

Gerber 1898
Gerber, Paul. "Die Räumliche und zeitliche Ausbreitung der Gravitation." *Zeitschrift für Mathematik und Physik* 43 (1898): 93–104.

Gerber 1917
Gerber, Paul. "Die Fortpflanzungsgeschwindigkeit der Gravitation." *Annalen der Physik* 52 (1917): 415–44.

Gibbon 1761
Gibbon, Edward. *Essai sur l'étude de la littérature.* London, 1761.

Ginzel 1899
Ginzel, F.K. *Spezieller Kanon der Sonnen- und Mondfinsternisse.* Berlin, 1899.

Ginzel 1906/1911/1914
Ginzel, F.K. *Handbuch der mathematischen und technischen Chronologie.* 3 vols. Leipzig, Germany: J.C. Hinrichshe Buchandlung, 1906, 1911, 1914.

Ginzel and Wilkens 1908
Ginzel, F.K., and A. Wilkens. *Theorie der Finsternisse.* Encyklopedie der Wissenschaftten, 1908.

Gleick 2003
Gleick, James. *Isaac Newton.* New York: Vintage Books, 2003.

Grafton 1975
Grafton, Anthony. "Joseph Scaliger and Historical Chronology: The Rise and Fall of a Discipline." *History and Theory* 14, 2 (1975): 156–85.

Grafton 1983, 1993
Grafton, Anthony. *Joseph Scaliger: A Study in the History of Classical Scholarship.* Vol. 1: *Textual Criticism and Exegesis.* Vol. 2: *Historical Chronology.* Oxford: Oxford University Press, 1983, 1993.

Grafton 2002
Grafton, Anthony. "A Premature Autobiography?" Unpublished manuscript, 2002.

Grafton 2003
Grafton, Anthony. "Some Uses of Eclipses in Early Modern Chronology." *Journal of the History of Ideas* 64 (April 2003): 213–29.

Graßhoff 1990
Graßhoff, Gerd. *The History of Ptolemy's Star Catalogue.* London: Springer Verlag, 1990.

Hardouin 1729
Hardouin, Père Jean. "Le Fondement de la Chronologie de M. Newton, Anglois, imprimé à Londre en 1726, sappé par le P. J. H." In *Mémoires de Trévoux*, September 1729, 1563–86.

Harpur 2004
Harpur, Tom. *The Pagan Christ.* Toronto: Thomas Allen, 2004.

Hay 1977
Hay, Denys. *Annalists and Historians: Western Historiography from the Eighth to the Eighteenth Centuries.* London: Methuen, 1977.

Herodotus 1904
Herodotus. *Histories.* London: G. Bell, 1904.

Hesiod 8 BC
Hesiod. *Works and Days.* 832 lines.

Hume 1902
Hume, David. *An Enquiry Concerning Human Understanding*, 114–16. Ed. L.A. Selby Bigge. Oxford: Clarendon Press, 1902.

James 1991
James, Peter. *Centuries of Darkness: A Challenge to the Conventional Chronology of Old World Archeology.* London: Jonathan Cape, 1991.

Jonsson 1986
Jonsson, Carl Olof. *The Gentile Times Reconsidered.* Atlanta: Commentary Press, 1986.

Kasparov 2002
Kasparov, Garry. "Mathematics of the Past." *Pi in the Sky* 5 (September 2002): 5–8.

Keller 1995
Keller, Werner. *The Bible as History.* New York: Barnes and Noble Books, 1995.

Kitchen 1983
Kitchen, Kenneth Anderson. *The Third Intermediate Period in Egypt (1100–650 BC).* Warminster, England: Airs & Phillips, 1973.

Kosambi 1965
Kosambi, D.D. *The Culture and Civilization of Ancient India in Historical Outline.* London: Routledge & Kegan Paul, 1965.

Larrabee 1950
Larrabee, Eric. "The Day the Sun Stood Still." *Harper's Magazine*, January 1950, 19–26.

Laskar 2003
Laskar, Jacques. "Immanuel Velikovsky: Monde en collision." *La Recherche* 368 (2003): 101.

Lehmann 1977
Lehmann, Johannes. *The Hittites: People of a Thousand Gods.* London: Collins, 1977.

Livy 1890–1911
Livy. *The History of Rome (Ad urbe condita . . .).* London: G. Bell, 1890–1911.

Loomis 1916
Loomis, Louise Ropes, trans. and ed. *The Book of Popes (Liber Pontificalis).* New York: Columbia University Press, 1916.

Lupoato 1997
Lupoato, Giovanni. *SN 1054: Una Supernova sul Medioevo.* Roma, 1997.

Manuel 1963
Manuel, Frank E. *Isaac Newton: Historian.* Cambridge, Mass.: Belknap Press of Harvard University, 1963.

Mencken 1937
Mencken, Johann Burkhard. *The Charlatanry of the Learned (De charlataneria eruditorum, 1715).* Translated from the German by Francis E. Litz, with notes and an introduction by H.L. Mencken. New York and London: A.A. Knopf, 1937.

Mommsen 1883
Mommsen, August. *Untersuchungen über das Kalenderwesen der Griechen, insbesonderheit der Athener.* Leipzig, Germany: B.G. Teubner Verlag, 1883.

Mommsen 1858
Mommsen, Theodor. *Die Römische Chronologie bis auf Caesar.* Berlin: 2. Auflage, Weidmansche Buchandlung, 1858.

Morozov 1997/98
Morozov, Nicolai A. *Christ: The History of Human Culture from the*

Standpoint of the Natural Sciences. 2nd ed. Moskow: Kraft & Lean, 1997/98 (in Russian).

Neugebauer, Parker, and Pingree 1982
Neugebauer, O., R.A. Parker, and D. Pingree. "The Zodiac Ceiling of Petosiris and Petubastis." In *Denkmäler der Oase Dachla: Aus dem Nachlass von Ahmed Fakhry. Archäologische Veröffentlichungen* 28, Deutsches Archäologisches Institut, Abteilung Kairo. Mainz, West Germany: Verlag Philipp von Zabern, 1982.

Neugebauer and Van Hosen 1959
Neugebauer, O., and H.B. Van Hosen. *Greek Horoscopes.* Philadelphia: The American Philosophical Society, 1959.

Newton 1728a
Newton, Isaac. *A Short Chronicle from the First Memory of Things in Europe to the Conquest of Persia by Alexander the Great.* London, 1728. Also published in the first edition of 1988. Complete text available at http://www.pereplet.ru/gorm/fomenko/inewton.htm

Newton 1728b
Newton, Isaac. *The Chronology of Ancient Kingdoms Amended.* London, 1728.

Newton MS (K)
Newton, Isaac. Keynes Manuscripts. King's College Library, Cambridge, England.

Newton MS (NC)
Newton, Isaac. New College Manuscripts. New College, Cambridge, England.

Newton 1977
Newton, Robert R. *The Crime of Claudius Ptolemy.* Baltimore: Johns Hopkins University Press, 1977.

Newton 1979, 1984
Newton, Robert R. *The Moon's Acceleration and Its Physical Origins.* Vol. 1: *As Deduced from Solar Eclipses.* Vol. 2: *As Deduced from General Lunar Observations.* Baltimore: Johns Hopkins University Press, 1979, 1984.

Nihongi 1956
Nihongi: Chronicles of Japan from the Earliest Times to AD 697. Trans. W.G. Ashton, 1896. London: Allen & Unwin, 1956.

Nosovski 1994
Nosovski, Gleb V. "The Dating of the First Ecumenical Council of Nicaea and the Beginning of the Christian Era." Appendix in Fomenko 1994, vol. 2, 390–411.

Nosovski 2003
Nosovski, Gleb V. *New Dating of Nativity and Crucifixion.* Isle of Man: Delamere Resources, 2003.

Novikov 2000
Novikov, Sergei P. "Pseudohistory and Pseudomathematics: Fantasy in Our Life." *Russian Mathematical Surveys* 55 (2000): 365–68.

Oppolzer 1962
von Oppolzer, Theodor Ritter. *Canon of Eclipses.* New York: Dover, 1962.

Payne-Gaposchkin 1950a
Payne-Gaposchkin, Cecilia. "Nonsense, Dr. Velikovsky." *Reporter*, March 14, 1950, 37–40.

Payne-Gaposchkin 1950b
Payne-Gaposchkin, Cecilia. "Nonsense, Dr. Velikovsky." *Popular Astronomy*, June 1950, 278–86.

Payne-Gaposchkin 1952
Payne-Gaposchkin, Cecilia. "Worlds in Collision." *Proceedings of the American Philosophical Society* 96 (1952): 519–25.

Peters and Knobel 1915
Peters, Christian Heinrich Friedrich, and Edward Ball Knobel. *Ptolemy's Catalogue of Stars: A Revision of the Almagest.* Washington, DC: The Carnegie Institution of Washington, Publication No. 86, 1915.

Plutarch 1928
Plutarch. *The Lives of the Noble Grecians and Romans.* Boston: Houghton Mifflin, 1928.

Reese, Craun, and Mason 1983
Reese, R.L., E.D. Craun, and C.W. Mason. "Twelfth-Century Origins of the 7980-Year Julian Period." *American Journal of Physics* 51 (1983): 73.

Reese, Everett, and Craun 1981
Reese, R.L., S.M. Everett, and E.D. Craun. "The Origin of the 7980-Year Julian Period: An Application of Congruences and the Chinese Remainder Theorem." *American Journal of Physics* 49 (1981): 658–61.

Reid 1728
Reid, Andrew. "The Chronology of Ancient Kingdoms Amended by Sir Isaac Newton." In *The Present State of the Republick of Letters*, vol. 2. London, 1728.

Reingold and Dershowitz 2001
Reingold, Edward M., and Nachum Dershowitz. *Calendrical Calculations: The Millennium Edition.* Cambridge: Cambridge University Press, 2001.

Repcheck 2003
Repcheck, Jack. *The Man Who Found Time: James Hutton and the Discovery of Earth's Antiquity.* Cambridge, Mass.: Perseus Publishing, 2003.

Rohl 1995
Rohl, David M. *A Test of Time: The Bible—from Myth to History.* London: Century, 1995.

Saari 1971–73
Saari, Donald G. "Improbability of Collisions in Newtonian Gravitational Systems." *Transactions of the American Mathematical Society* 162 (1971): 267–71; 168 (1972): 521; 181 (1973): 351–68.

Saari 1975
Saari, Donald G. "Collisions Are of First Category." *Proceedings of the American Mathematical Society* 47 (1975): 442–45.

Samuel 1972
Samuel, Alan E. *Greek and Roman Chronology: Calendars and Years in Classical Antiquity.* München: C.H. Beck'sche Verlagsbuchandlung, 1972.

Scaliger 1927
Scaliger, Joseph Justus. *Autobiography.* Cambridge, Mass.: Harvard University Press, 1927.

Schimmelpfennig 1992
Schimmelpfennig, Bernhard. *The Papacy.* New York: Columbia University Press, 1992.

Schove 1984
Schove, Justin D. *Chronology of Eclipses and Comets, AD 1–1000.* Woodbridge, Suffolk: Boydell Press, 1984.

Sheynin 1993
Sheynin, Oscar. "The Treatment of Observations in Early Astronomy." *Archive for History of Exact Sciences* 46, 2 (1993): 153–92.

Spence 1820
Spence, Joseph. *Observations, Anecdotes, and Characters of Books and Men.* London, 1820.

Stephenson and Fatoohi 2001
Stephenson, F. Richard, and Louay J. Fatoohi. "The Eclipses Recorded by Thucydides." *Historia* L/2 (2001): 245–53.

Stewart 1951
Stewart, John Q. "Disciplines in Collision." *Harper's Magazine*, June 1951, 57–63.

Strassler 1996
Strassler, Robert B., ed. *The Landmark Thucydides: A Comprehensive Guide to the Peloponnesian War.* New York: The Free Press, 1996.

Tacitus 1907
Tacitus, Cornelius. *The Annals.* London: Oxford University Press, 1907.

Taylor 1987
Taylor, R.E. *Radiocarbon Dating: An Archeological Perspective.* New York: Academic Press, 1987.

Taylor 2000
Taylor, Timothy. "Time Warp." *Saturday Night*, September 16, 2000, 39–45.

Thucydides 1849
Thucydides. *The History of the Peloponnesian War.* London: H.G. Bohn, 1849.

Toomer 1984
Toomer, G.J. *Ptolemy's Almagest.* New York: Springer Verlag, 1984.

Toynbee 1934
Toynbee, Arnold Joseph. *A Study of History.* London: Oxford University Press, 1934.

Twain 1961
Twain, Mark. *The Autobiography of Mark Twain.* New York: Washington Square Press, 1961.

Velikovsky 1950
Velikovsky, Immanuel. *Worlds in Collision.* New York: Doubleday, 1950.

Velikovsky 1951a
Velikovsky, Immanuel. "Answer to My Critics." *Harper's Magazine*, June 1951, 51–57.

Velikovsky 1951b
Velikovsky, Immanuel. "Answer to Professor Stewart." *Harper's Magazine*, June 1951, 63–66.

Velikovsky 1952
Velikovsky, Immanuel. *Ages in Chaos.* New York: Doubleday, 1952.

Wachsmuth 1895
Wachsmuth, Curt. *Einleitung in das Studium der Alten Geschichte.* Leipzig, Germany, 1895.

Waerden 1988
Waerden, Bartel Leendert van der. *Die Astronomie der Griechen.* Darmstadt, West Germany: Wissentschaftliche Buchgesellschaft, 1988.

Walton 2004
Walton, Dawn. "Canadian Archaeologists Shaking Up Nicaragua." *Globe and Mail* (Toronto), August 27, 2004, A5.

Wantzel 1837
Wantzel, Pierre L. "Recherches sur les Moyens de Reconnaître si un Problème de Géometrie peut se Résoudre avec la Règle et le Compas." *Journal de Mathématiques* 2 (1837): 366–72.

Westfall 1980
Westfall, Richard S. *Never at Rest: A Biography of Isaac Newton.* New York: Cambridge University Press, 1980.

Westfall 1993
Westfall, Richard S. *The Life of Isaac Newton.* New York: Cambridge University Press, 1993.

Wilson 2000
Wilson, Ian. *The Bible Is History.* Washington, DC: Regnery Publishing, 2000.

Zalyzniak 2000
Zalyzniak, A.A. "Linguistics According to A.T. Fomenko." *Russian Mathematical Surveys* 55 (2000): 369–404.

Acknowledgments

Most of this book was written at home in Victoria, B.C., but some parts of it took shape while I attended conferences or made research trips to Banff (Canada), Beijing (China), Tokyo (Japan), Sofia (Bulgaria), Sibiu (Romania), Cala Gonone (Sardinia, Italy), Cargèse (Corsica, France), Lausanne (Switzerland), and La Serena and Santiago (Chile).

I first of all want to thank Mariana Diacu for reading the first draft of every chapter and for making excellent suggestions and remarks; I aimed this book at smart non-expert readers like her, and her feedback was extremely useful. I am also very indebted to my publisher, Diane Martin, whose intelligence, enthusiasm, and vision have guided my later drafts; to Louise Dennys, who supported my proposal from the very beginning; to my editors, John Eerkes-Medrano, Rosemary Shipton, and John Sweet for their excellent contributions to the manuscript; and to my agent, Kathryn Mulders, without whom I might never have started on this project.

I also want to thank the many other people who helped me in various ways, such as reading parts of the manuscript and commenting on them or providing me with information and suggesting references. They are Marco Abate, Elizabeth von Aderkas, Vicky Alfred, Arif Babul, Victor Planas Bielsa, Wendy Brandts, Crystal Brasseur, Alex Buium,

Mortimer Chambers, Susan Chen, Adrian Damian, Chandler Davis, Douglas Donahue, Ivar Ekeland, Gordon Emslie, Maggi Feehan, Jim Ferguson, Anatoli Fomenko, Jeff Foss, Rick Gibbs, Margaret Gracie, Anthony Grafton, Dieter Hagedorn, Kitty Hoffman, Paul Hoffman, Philip Holmes, Eva Horvath, Andrei Iacob, Wali Kameshwar, Gary Kasparov, Vikie Kearn, Wieslaw Krawcewicz, Jacques Laskar, Trevor Lipscombe, Ellen Maddow, Jerry Marsden, Vasile Mioc, Bob Moody, Gleb Nosovski, Liviu Ornea, Robert Osserman, Osmo Pekonen, Ernesto Pérez-Chavela, Florian Pop, Tudor Ratiu, Greg Rowe, Don Saari, Alan Samuel, Marjorie Senechal, Greg Sieb, F. Richard Stephenson, Kyotaka Tanikawa, Timothy Taylor, Min Tsao, Dianne Tyers, and Jan Zwicky.

Last but not least, I want to thank the dedicated staff at Knopf Canada who produced and advertised this book: Frances Bedford, Nina Ber-Donkor, Angelika Glover, Janine Laporte, Deirdre Molina, Scott Richardson, and Leah Springate. It was a pleasure to work with all of them.

Finally, I would like to apologize to anyone who helped me with this book but whom I inadvertently omitted to mention.

Index

FLORIN DIACU is a Professor of Mathematics and former Director of the Pacific Institute for the Mathematical Sciences at the University of Victoria. He is the author of *Celestial Encounters*, a history of ideas in the field of chaos theory. He lives in Victoria, British Columbia.

A NOTE ABOUT THE TYPE

The Lost Millennium is set in a digitized form of Times New Roman. Designed by The Monotype Corporation as a specific commission for *The Times of London* in the early 1930s, the typeface was required to be not just a general purpose font family but one with strong lines, firm contours and above all, economy of space. Research into legibility and readability resulted in a face unique to newspaper typography of the time, one that was more condensed and featured greater contrasts than others.

Times New Roman continues to be one the world's most widely used typefaces, particularly in print media and academic publishing.